MIRAGE OF HEALTH

Utopias, Progress, and
Biological Change

BY RENÉ DUBOS

Anchor Books
Doubleday & Company, Inc.
Garden City, New York

CONTENTS

MIRAGE OF HEALTH

Utopias, Progress, and Biological Change

It seems to me, that in living so far, through all our bitter centuries of civilization, we have still been living onwards, forwards. . . . The past, the Golden Age of the past—what a nostalgia we all feel for it. Yet we don't want it when we get it.

D. H. LAWRENCE

I.

THE GARDENS OF EDEN

The Golden Ages

Belief in a golden age has provided mankind with solace in times of despair and with élan during the expansive periods of history. Dreamers imagine the golden age in the remote past, in a paradise lost, free from toil and from grief. Optimists put their faith in the future and believe that mankind, Prometheus-like, will master the arts of life through power and knowledge. Thus, the golden age means different things to different men, but the very belief in its existence implies the conviction that perfect health and happiness are birthrights of men. Yet, in reality, complete freedom from disease and from struggle is almost incompatible with the process of living.

Life is an adventure in a world where nothing is static; where unpredictable and ill-understood events constitute dangers that must be overcome, often blindly and at great cost; where man himself, like the sorcerer's apprentice, has set in motion forces that are potentially destructive and may someday escape his control. Every manifestation of existence is a response to stimuli and challenges, each of which constitutes a threat if not adequately dealt with. The very process of living is a continual interplay between the individual and his environment, often taking the form of a struggle resulting in injury or disease. The more creative the individual the less he can hope to avoid danger, for the stuff of creation is made up of responses to the forces that impinge on his body and soul. Complete and lasting freedom from disease is but a dream remembered from imaginings of a Garden of Eden designed for the welfare of man

The illusion that perfect health and happiness are within man's possibilities has flourished in many different forms throughout history. Primitive religions and folklores are wont to place in the remote past this idyllic state of paradise on earth, most ancient peoples have in their legends stories of happier times, when men enjoyed long lives during which they remained strong and healthy. In the Old Testament the Patriarchs are said to have lived hundreds of years, while their descendants can hardly aspire to more than threescore and ten. The ancient Greeks believed in the existence of happy races, vigorous and virtuous, in inaccessible parts of the earth. According to their legends, the Hyperboreans and the Scythians in the north, the Ethiopians in the south, lived exempt from toil and warfare, from disease and old age, in everlasting bliss like the dwellers in the Isles of the Blest at the edge of the Western Sea. In *Works and Days* Hesiod wrote of the golden age when men "feasted gaily, undarkened by sufferings" and "died as if falling asleep." The oldest known medical treatise written in the Chinese language also refers to the health of the happy past. "In Ancient times," states the Yellow Emperor in his *Classic of Internal Medicine* published in the fourth century B C., "people lived to a hundred years, and yet remained active and did not become decrepit in their activities. . . . But eventually the tranquil era came to an end, and as men turned more violent they became more vulnerable to noxious influences."[1]

While the events that brought to an end the legendary era of health and happiness were placed in distant countries by the Greeks and in the remote past by the Chinese, the violent changes responsible for increase in emotional and physiological misery are not always vague beliefs arising from the mists of time. For a few peoples, indeed, they

[1] See for example Veith Ilza, *Huang Ti Nei Ching Su Wen, The Yellow Emperor's Classic of Internal Medicine*, Williams & Wilkins, Baltimore, 1949, 253 pp.

are the precise and well-documented memories of recent disasters.

It was as recently as 1864, for example, that the Navajo Indians were overrun by Kit Carson in Canyon de Chelly in Arizona. The destruction of their gardens and peach trees in the innermost holy land of the tribe broke their spirit and terminated their resistance. Twenty-five thousand of "The People" made the long walk to Fort Sumner in eastern New Mexico, where they were held in captivity until 1868. That year some ten thousand of the survivors were allowed to return and settled in the desertic territory which now constitutes the Navajo reservation. There the tribe succeeded in adapting itself to a peaceful pastoral way of life based on the exploitation of sheep and goats. Living on a diet high in meat products, supplemented with Indian corn, wild berries, fruits and nuts, the Navajos rapidly increased in numbers despite great physical hardships and high infant mortality. Today, their population exceeds eighty thousand.

Some two decades ago, however, the Navajo tribe experienced new trials that disorganized its economy. In an attempt to control the erosion brought about by overgrazing and to prevent the irremediable loss of pasture lands, the Indian Bureau of the federal government decided in 1934 to limit drastically the number of animals grazed on the reservation. As a result of this administrative measure, useful in the long run and indeed essential, the Navajo people were compelled to limit the size of their herds, and in consequence had to shift suddenly from a way of life based on the products of grazing to one in which most of the necessities had to be obtained at the trading post. The basic staples of Navajo life which had consisted of meat, milk, goatskins, and wool were replaced within a few years by white flour, molasses, canned food, soda drinks, and cheap manufactured goods. This profound and sudden change in tribal economy brought in its train much mental and physical hardship, expressing itself in various forms of disease. Even today many Navajos still regard the sheep-reduction program as marking the end of happy times and the origin

of their trials. A few years ago two of their most trusted
white friends translated in the following words some of the
pathetic laments that were told them on the reservation:

> A long time ago, before we were born, the white
> people and our old folks made a treaty This treaty was
> made to the end that these Encircling Mountains
> would always be ours, so that we could live according
> to them. The right to these was given to us, so that
> all the Navajos might live in accord with that which
> is called Mountain Soil, and the pollen of all the plants.
> All Navajos live in accord with them. With these sacred
> things everything was stabilized and our wealth in-
> creased, but then the white man took them.
>
> All these things that we once raised have been taken
> from us. The sheep and the goatskin robes are gone;
> the thick fleeces used as bedding are gone. . . . Long
> ago children with tuberculosis were unheard of. But
> when things like the goatskin robes passed out of ex-
> istence one began to hear about many children who
> were sick with tuberculosis and other diseases.
>
> The womenfolk and children wept for their goats.
> The bleating of the milk goats that fed the children
> faded away into the distance, and the wails of the chil-
> dren arose in their stead—the wails of the children and
> the womenfolk. They did not weep without reason, for
> the food of the people had been taken from them.
>
> We suffered from everything, from hunger, from
> lack of meat and from despondency. The Special Graz-
> ing Regulation is like a killing disease from which one
> cannot sleep.[2]

What the sheep and goats used to be for the Navajos
the buffalo was for the Plains Indians and the caribou was
for the Eskimos of the Great Barrens. These animals symbolized the dependence of man upon his environment, as
corn still does today for the Pueblo Indians. Before they

[2] Robert W. Young and William Morgan, *Navajo Historical
Selections*, Navajo Historical Series No. 3. Bureau of Indian
Affairs, 1954.

came into contact with the white man, many primitive people all over the world had developed to an extraordinary degree the art of making use of the natural resources available to them from the land, the water, and the air. Through ancestral skills, customs, and taboos they had learned to cope with most of the physical and biological dangers which threatened their existence. In brief, they had achieved all sorts of subtle adaptations to their total environment which permitted them to survive even under the most adverse circumstances and to achieve relative health and happiness, at least for limited periods. Like many other primitive peoples, the Navajos achieved harmonious equilibrium with nature by living in accord "with the mountain soil, the pollen of the native plants, and all other sacred things." In their tribal memory the freedom to roam with sheep and goats over the Arizona desert without government control became transmuted into the idyllic happiness of their golden age.

The Return to Nature

Like primitive peoples, men in civilized societies commonly believe in the possibility of an ideal state of health and happiness. But, instead of expressing this belief through legends and folklore, they are apt to rationalize it in the form of philosophical theories and to assert that a healthy mind in a healthy body can be achieved only by harmonizing life with the ways of nature. "The deviation of Man from the state in which he was originally placed by Nature seems to have proved to him a prolific source of Disease," wrote Jenner in the introduction to his famous essay, *The Cow Pox*, by which he introduced to the world the not-so-natural practice of vaccination. Jenner was not original in making this statement, for throughout the ages and all over the world it has been a common illusion that good life could be identified with natural life.

As a reaction against the pompous formality of the Grand Siècle, the latter part of the eighteenth century proved particularly receptive—in theory at least—to the gospel that all human problems could be solved by returning to the ways of nature. Almost everyone but Voltaire listened with ecstasy when Jean Jacques Rousseau asserted that man in his original state was good, healthy, and happy and that all his troubles came from the fact that civilization had spoiled him physically and corrupted him mentally. *"Tout est bien sortant des mains de l'auteur des choses,"* wrote Rousseau, *"tout dégénère entre les mains de l'homme."* The ideal man was therefore the savage, untainted by civilization; the ideal life demanded direct communion with nature and independence from conventions. "Hygiene," Rousseau claimed, "is less a science than a virtue," and at the turn of the century Thomas Beddoes echoed this attitude in the revealing title of his book *Hygeia, or Essays Moral and Medical on the Causes Affecting the Personal State of our Middling and Affluent Classes.* Sickness being the result of straying away from the natural environment, the blessed original state of health and happiness could be recaptured only through abiding by the simple order and purity of nature—or, as Voltaire said in maliciously paraphrasing Rousseau, through learning again to walk on all fours.

Rousseau's sentimental message, if neither original nor profound, was at least eloquent and timely. To a world satiated with the refinements of the eighteenth-century civilization it brought the naive but attractive picture of a life in which men would be free of vices and of physical ailments because free of unnatural wants and worry—as was assumed to be the case for primitive peoples. Even the most uncritical followers of Rousseau acknowledged, of course, that sophisticated Europeans could hardly be brought back to a state compatible with primitive life—happy and healthy as it might prove to be. In *L'Ingénu* Voltaire made much fun of the theme by describing the behavior of a Huron Indian in France. Despite the simple manners of the young savage, the uninhibited play of his natural appetites and instincts led to extremely embarrassing situations in the social milieu of Europe.

While the charm of primitive life was much talked about in the salons, it is not apparent that many literary Europeans

seriously considered emigrating to the American forests or
to the South Sea Islands. Instead they imagined or pretended that the values of primitive life could be recaptured
by modifying the physical aspect of Europe to make it more
nature-like. Under the influence of John Locke, and particularly of Horace Walpole, the English landscape archi-
tects had already begun to develop a new type of scenery
which they thought natural because free of geometrical de-
sign Their ideal was to compose the landscape not by artificial rules but in accordance with the topographical and
other natural peculiarities of the place. Their guide was
Pope's admonition to "consult the genius of the place in all."
In order to bring out the true "genius" of English scenery,
Capability Brown, Humphrey Repton, and their followers
ruthlessly destroyed the beautiful classical gardens that had
graced Tudor life and, for the sake of informality, allowed
cattle to graze along synthetic serpentine rivers in view of
the noble English mansions. Even the French fell under the
sway of this doctrine and came to scorn the formal magnifi-
cence of their parks and gardens. They imitated the semi-
natural style of the English gardenists and landscape archi-
tects despite Horace Walpole's warning that France could
never match the luxuriance of English scenery because of
lack of verdure and of water. "They can never have as
beautiful landscapes as ours," he wrote, "till they have as
bad a climate." Carrying the fashionable craze for nature
to an absurd but charming extreme of logic Marie Antoi-
nette built on the edge of the park of Versailles her synthetic Hameau where powdered marchionesses played with
her at haymaking and tending cows, thereby pretending
to experience the idyllic simplicity of rustic life.

Let it be mentioned here in passing that the glorification
of the noble savage and of his natural life did not originate
during the eighteenth century nor with Rousseau's writings.
Since very ancient times the theory that most of the ills of
mankind arise from failure to follow the laws of nature has
been endlessly reformulated in every possible form and
mood, in technical and poetical language, in ponderous
treatises and witty epigrams. As already mentioned, it was

that will-of-the-wisp the "golden age" which had inspired
Hesiod in Greece and the medical writings of the Yellow
Emperor in ancient China. In particular, the Taoist philosophy which has so profoundly influenced Chinese life and
art is pervaded by reverence for nature. In the *Tao Tê
Ching* (*The Way*) Lao-tzu wrote, probably before 500
B.C., "Have you ever heard of the Age of Perfect Character?
In the old days the people tied knots for reckoning. They
enjoyed their food, beautified their clothing, were satisfied
with their homes, and delighted in their customs. Neighboring settlements overlooked one another, so that they would
hear the barking of dogs and crowing cocks of their neigh-
bors, and the people till the end of their days had never
been outside their own country In these days there was
indeed perfect peace," Lao-tzu, the Jean Jacques Rousseau
of ancient China, was followed by many translators and
imitators who restated his message that man must merge
himself with his surroundings and move along with them.
Chuang-tzu wrote of the time when "the ancient men lived
in a world of primitive simplicity. . . . That was the time
when the *yin* and the *yang* worked harmoniously, and the
spirits of men and beasts did not interfere with the life of
the people, when the four seasons were in order and all
creation was unharmed, and the people did not die young."

An appealing vision of the Taoist paradise on earth
emerges from the writings of Lieh-Tzu (fifth to third century B.C.). In the happy land that he describes, "The people were gentle, following Nature without wrangling and
strife. . . . Men and women wandered freely about in company; marriage-plans and betrothals were unknown. Living
on the banks of the rivers, they neither ploughed nor har-
vested, and since the *chhi* of the earth was warm, they
had no need of woven stuffs with which to clothe themselves. Not till the age of a hundred did they die, and disease and premature death were unknown. Thus they lived
in joy and bliss, having no private property; in goodness
and happiness, having no decay and old age, no sadness or
bitterness." At a later date the radical thinker Pao Ching-
yen evoked again the ancient times when

there were no lords and officials. . . . Man in the
morning went forth to his labour on his own accord
and rested in the evening. People were free and uninhibited and at peace; they did not compete with one
another, and knew neither shame nor honours. There
were no paths on the mountains, and no bridges over
waters nor boats upon them, nor were the rivers made
navigable. Thus invasions and annexations were not
possible. The myriad beings participated in a mysterious equality and forgot themselves in the Tao. Con-
tagious diseases did not spread, and long life was followed by natural death. The hearts of men were pure
and innocent of ruses and deceits. Having enough to
eat, the people were contented, patted themselves on
the belly, and wandered about for pleasure.[3]

In Europe the era of great maritime explorations brought
first the navigators, then missionaries, soldiers, and merchants, in actual contact with primitive cultures all over the
world. The appealing character of uncivilized people entered European consciousness through the semifactual accounts of De Léry's *Voyage au Brésil*, published in 1556–
58. One century later an engaging picture of primitive life
appeared in the description of Eskimo life by Nicolas
Tunnes, who had observed it in West Greenland in 1656.
He wrote:

Although they [the Eskimos] are one of the poorest
and most barbarous nations under the sun, they believe themselves to be very happy, and the best favored people in the world. . . . They know nothing of
all those gnawing cares and besetting sorrows which
torment most other people. . . . All their efforts are di-
rected toward acquiring, without too much trouble,
what is absolutely necessary in the way of clothing and
food. . . . They eat all their food without cooking it,

[3] See for example Joseph Needham, *Science and Civilisation
in China*, Vol. 2, *History of Scientific Thought*, University Press,
Cambridge, 1956.

and with no other sauce than that supplied by their keen appetite.

More than anything else, however, it was the reports on the "Islands of Paradise" by the explorers of the South Pacific which convinced Europe that carefree life and free love, untrammeled by the complexities of civilization, were the happy lot of the "natural man." In 1766 Captain Samuel Wallis set sail from Plymouth in a 511-ton, copper-bottomed ship called the *Dolphin* and after a dangerous voyage discovered the enchanted land of Tahiti in June 1767. Following a few skirmishes in which the ship replied with musket fire and grapeshot to the stones of the natives, hostilities came to an end. The weary sailors found the island a paradise where the scenic beauty was enhanced by the softness of the climate and the amorous welcome of the women. Wallis had done more than discover a convenient port of call in Tahiti, for Polynesian life was to become the symbol of a new attitude toward nature and a fountain of inspiration for the romantic movement.

Shortly after Wallis, Louis de Bougainville also reached Tahiti in his ship *La Boudeuse* The natives were as friendly to the French explorers, and the women as kind, as they had been to Wallis' men. Philibert Commerson, the physician and naturalist who accompanied Bougainville on the *Boudeuse*, published a highly romanticized account of the Tahitians, even regarding their skill in thievery as a sort of primitive communism. He wanted Tahiti called Utopia or The Fortunate Isle, because he saw in it the true and factual embodiment of Thomas More's dreams. But Bougainville, more poetical, had already named the enchanted isle La Nouvelle Cythère!

Most influential, probably, was the voyage that Captain James Cook made on the *Endeavour* through the South Pacific between 1768 and 1771. Just as much as Wallis and Bougainville, the sober, disciplined Captain Cook, genius of the matter-of-fact, was obviously captivated by the charm of the Pacific Islands and of Polynesian life. "The whole scene realized the poetical fables of Arcadia," he wrote in his *Journal,* and his admiration extended to the vigor, physical beauty, and healthy appearance of men and women, as well as to the air of happiness that prevailed through the islands. More unexpected is the flattering account that Cook gave of the Australian aborigines with whom he first came into contact in 1771. "They may appear to some to be the most wretched people upon earth, but in reality they are far more happy than we Europeans. . . . They live in a tranquillity which is not disturbed by the inequality of condition. The earth and sea of their own accord furnishes them with all things necessary for life. . . . They have very little need of clothing, and this they seemed fully sensible of, for many to whom we gave cloth left it carefully upon the sea beach and in the woods."

Throughout the eighteenth century the accounts of the explorers, as well as the writings of their popularizers and rehashers, were read widely by an avid public. The progress of Cook's ship was covered in the English newspapers of the time with tricks of reporting and journalistic style worthy of the twentieth-century press. *Hawkesworth's Voyages*—a semiofficial account of the "Discoveries in the Southern Hemisphere . . . drawn up from the Journals which were kept by the several Commanders" had two editions in London in 1773 and a third in 1785. A New York edition and French and German translations appeared in 1774 and an Italian one in 1794. Translations in other languages continued during the early part of the following century. In war-torn Europe, the descriptions of the simple life of primitive people appeared like balm from an exotic world and the flood of books on exploration catered to the taste of the time by extolling the health, virtues, and happiness of the natives. In his *Observations* Forester gave a "General View of the Happiness of the Islanders in the South Sea." The highest degree of happiness, according to him, was in Tahiti.

There was probably much truth in the opinion widely held in the late eighteenth century that the Polynesians were then like Stone Age aristocrats. Many of them had a magnificent physical development and a happy carefree life, even though all were on the edge of poverty and were overcrowded according to European standards. But it is also certain that everything was not so idyllic in the Polynesian world as reported by the navigators. In most Pacific islands the social structure was based on the conquest of the original inhabitants by a later arriving ruling clan. The newcomers in the islands had killed or absorbed the relatively small number of aborigines and had proliferated their own race, but in other islands the result of conquest was a ruling class and a ruled. In Tahiti, as elsewhere, it was chiefly the ruling class which provided men of tall, impressive mien. There is also good evidence that the sexual freedom which so much impressed the voyagers was less widespread than they believed and that Polynesian aristocracy had more exclusive standards than the socially inferior. The fleeting visitor probably failed to recognize that the European gadgets which he brought as gifts had a special appeal for a rather limited group of men and women who could easily be bribed. He had little occasion to suffer from the stratification of society or to observe certain practices, including human sacrifice, which would have shocked him and dispelled some of his illusions.

Cook, his contemporaries, and their romantic followers failed to notice the defects of primitive life probably because of the climate of opinion then prevailing in Europe —chiefly as a result of the magic of Rousseau's message. Literature had created the myth of the noble savage, and the myth had spread through all social classes. Little surprise that sailors aching with scurvy, fed on salt beef and hard biscuit, and sexually repressed should have thought themselves imparadised on reaching the balmy shores and the friendly people of the South Seas. So pervasive was the influence of the age that it could blind the judgment even of hard-boiled sailors and of a man as objective and thoughtful as Captain Cook.

Whatever the factual accuracy of the reports of the explorers concerning the health and happiness of primitive people, they provided flesh and substance to Rousseau's glorification of the noble savage. They fostered a state of mind which persisted long into the nineteenth century and found literary expression in the writings of Chateaubriand on the Natchez Indians and in Melville's account of the paradise he found in Typee. This attitude died progressively during the second half of the century, in part perhaps because Polynesians and American Indians had by that time become physically and mentally degraded through the venereal diseases, tuberculosis, acute infections, and alcoholism brought to them by the white sailors, soldiers, and merchants. After the white man had destroyed the valuable attributes of primitive cultures, his attitude became again one of contempt for the savage and he recovered assurance in the superiority of his own civilization.

Even after the cult of the noble savage had lost its appeal the Western world continued to hold to the belief that primitive man had once been close to the essence of things and had enjoyed health of body and mind. The conviction that man is healthy as long as he remains in intimate contact with "nature" appears even in as unlikely a place as De Quincey's *Confessions of an English Opium Eater*. "Opium," De Quincey wrote, "gives that sort of vital warmth which would probably always accompany a bodily constitution of primeval and antediluvian health." Much of the naturalist literature of the late nineteenth century also expresses a similar belief. Although it differs so profoundly in style from the elegiac writings of the eighteenth century, it shares with them a common basis of belief in the virtues of nature.

With less obvious affectation than the wigged and powdered shepherds and shepherdesses of the Petit Trianon, but also with less grace, modern man has done many odd things to display his faith in the fundamental goodness of nature. Following in the steps of Rousseau, one hundred million Central Europeans went botanizing in the hope of discovering among lowly flowers both the soul of the universe and natural remedies for chest troubles. More prosaic twentieth-century man tries to re-establish contact with this forgotten biological past in countless country clubs, hunting or ski lodges and beach bungalows, through clambakes in the moonlight and barbecue parties in suburban gardens and picnic groves. Nature cults and practices have sent people in all walks of life tramping barefoot in the morning dew, exposing themselves to discomfort in the wind and sun, drinking ill-tasting plant and animal juices, and imagining that the compost heap in the garden can become a fountain of youth. Whatever his inhibitions and tastes, Western man believes in the natural holiness of seminudism and raw vegetable juice, because these have become for him symbols of unadulterated nature.

Health through Science

In addition to its picturesque effects, the gospel of the return to nature stimulated during the Enlightenment a flood of technical books devoted to the preservation of health. In this case, again, the movement had begun long before the eighteenth century. Ever since Rabelais and Montaigne, the health and education of children had been much-discussed topics. John Locke had taught that the child was normally healthy because close to nature but soon lost his happiness through the fault of adults who, not understanding his needs, mismanaged his life and ruined his health. The theme, restated later and made fashionable by Rousseau in *Emile*, fired the world of the eighteenth century and found expression in a host of sentimental guidebooks on the proper manner of raising children. There is no doubt that this popular form of science, adorned with attractive pictures illustrating sensible ways to clothe young people at all stages of life, influenced vestimentary fashion. It is probable also that it contributed somewhat to increasing the comfort and happiness of the young during the succeeding generations.

This intellectual approach to the problems of health gave the illusion that in medicine, as in other social sciences, the Age of Reason would mark the beginning of a new era. In fact, there was justification for the optimism prevailing in medicine during the period 1750–1800. Leprosy and plague had all but disappeared from Europe; smallpox, malaria, and summer diarrhea had been brought under partial control. Condorcet envisaged an era in which man would be free from disease and old age and death would be indefinitely postponed; Benjamin Franklin made similar predictions. To achieve old age had a universal fascination and, then as now, there were probably many writers who pretended to believe that life begins at 40. *The Art of Prolonging Life* by C. W. Hopeland, Faust's *Catechism of Health,* and many similar medical treatises published in the eighteenth century were translated into all European languages and reprinted again and again. The most ambitious attempt to relate health to natural living was made by Johann Peter Frank with his *Mediziniche Polizei,* the nine volumes of which began to appear in 1779. In this ponderous treatise Frank formulated the thesis that illness was caused not only by physical factors but as much, if not more, by noxious influences originating in the social environment—from poverty to excessive love of the theater.

Problems of social medicine were much in the mind of the Encyclopedists, who formulated a scientific philosophy of public health which emphasized the complex relationships between social environment and the physical well-being of man. In 1820 the physician-philosopher Virey presented this subject in a book entitled *L'Hygiène Philosophique.* Virey believed that man in the state of nature possesses an instinct of health that permits him to adapt his biological behavior to the resources and dangers of his environment, just as wild animals do. But, wrote Virey, civilized man has lost this instinct of health and it is now the business of science to rediscover for him in the form of exact knowledge the biological wisdom that once was his birthright. According to the Encyclopedists, science not only was an instrument of progress but would soon bring about the millennium. After Charles Darwin man no longer looked back with regret upon a Paradise Lost, but instead began to think that he was approaching the gates of a new Eden. The golden age of humanity which had so long been placed in the inaccessible past by poets and romantic philosophers was now promised by scientists for the near future.

As we shall see, however, and general opinion notwithstanding, it is not exact laboratory science that has provided modern man with the best substitute for the instinct of health postulated by Virey. The most effective techniques to avoid disease came out of the attempts to correct by social measures the injustices and the ugliness brought about by industrialization. The contrast between the conditions of civilized man and the idyllic natural state described by the discoverers of the South Pacific became particularly shocking with the advent of the Industrial Era. The huge populations suddenly crowded into the factories and tenements of the mushrooming cities lived in squalor and were exposed to great physical and emotional hardships. Their privations and physiological misery created everywhere social and health problems so acute that they became an obsession for the European conscience. As a result, reform movements started all over the Western world almost spontaneously and simultaneously, and their momentum increased during the second half of the nineteenth century

We shall consider in a later chapter how the concern with social reforms rapidly evolved into public health practices that brought about spectacular improvements in the sanitary and nutritional state of the Western world. Suffice it to state here that this achievement cannot be credited to the type of laboratory science with which we are familiar today. Rather, it was the expression of an attitude which is almost completely foreign to the modern laboratory scientist. The nineteenth-century reformers naively but firmly believed that, since disease always accompanied the want, dirt, pollution, and ugliness so common in the industrial world, health could be restored simply by bringing back to the multitudes pure air, pure water, pure food, and pleasant surroundings—the qualities of life in direct contact with nature. There is no doubt that this philosophy, unsophisticated in terms of modern science, was nevertheless immensely effective in overcoming many of the disease problems brought into being by the Industrial Revolution. All contemporary observers expressed the view that the general conditions of health in Western Europe and in North America were much better during the second half of the nineteenth century than they had been before the social reformers began clearing up the mess caused by the sudden growth of industrial cities. And the improvement clearly began long before the modern era in medicine was ushered in by the germ theory of disease.

Although the laboratory scientist was only the laborer of the eleventh hour in the campaign against disease that began a century and a half ago, he occupies now the center of the stage everywhere, and practically all recent progress has been the result of his work. The nineteenth-century reformers had immense practical achievements to their credit, but the science of which they boasted was usually made up of catchwords. By preaching the virtues of pure air, pure water, and pure food they had gone far toward eliminating infection and improving nutrition, but their success had been due more to zeal in the correction of social evils than to understanding of medical problems. The original contribution of the laboratory scientist was to reformulate the problems of disease in more precise terms and to uncover by analysis the properties that lay hidden behind the word "pure"—so alluring yet so vague. And what the scientist uncovered was far more complex than the most subtle imaginings of all philosophers, humanitarians, and social reformers. He found among other things that poisons and germs of disease could lurk unseen in fragrant air and in limpid water and that the most tasty food, even though natural and pure, might be deficient in essential growth factors or be so unbalanced as to cause metabolic misery.

As the result of the scientist's labors, it became clear that the instincts of health postulated by Virey involved in final analysis all the complex and interrelated controls of physiological function. This new understanding, even though still so incomplete today, has yielded new and more convenient techniques for the control of some of man's ancient plagues. The time has passed when explorers on land or at sea have to depend on heavy loads of lemons and animal food in order to protect themselves against scurvy and other deficiency diseases. A few small packages of synthetic vitamins can now make an adequate diet out of proteins, carbohydrates, fats, and water. A dash of chlorine and an effective filtration bed will make any water supply more typhoidproof than the most sparkling streams brought from high mountains. There is no longer any reason to fear that bad air will kill men with yellow fever in the Brazilian jungle or with Carrion's disease in the Peruvian Andes. The knowledge that these diseases are caused by parasites transmitted through mosquitoes has led to protective measures far more effective than the traditional practices of natives in those regions.

But while modern science can boast of so many startling achievements in the health fields, its role has not been so unique and its effectiveness not so complete as is commonly claimed. In reality, as already stated, the monstrous specter of infection had become but an enfeebled shadow of its former self by the time serums, vaccines, and drugs became available to combat microbes. Indeed, many of the most terrifying microbial diseases—leprosy, plague, typhus, and the sweating sickness, for example—had all but disappeared from Europe long before the advent of the germ theory. Similarly, the general state of nutrition began to improve and the size of children in the labor classes to increase even before 1900 in most of Europe and North America. The change became noticeable long before calories, balanced diets, and vitamins had become the pride of nutrition experts, the obsession of mothers, and a source of large revenues to the manufacturers of colored packages for advertised food products.

Clearly, modern medical science has helped to clean up the mess created by urban and industrial civilization. However, by the time laboratory medicine came effectively into the picture the job had been carried far toward completion by the humanitarians and social reformers of the nineteenth century. Their romantic doctrine that nature is holy and healthful was scientifically naive but proved highly effective in dealing with the most important health problems of their age. When the tide is receding from the beach it is easy to have the illusion that one can empty the ocean by removing water with a pail. The tide of infectious and nutritional diseases was rapidly receding when the laboratory scientist moved into action at the end of the past century.

The great increase in over-all expectancy of life during the past hundred years in the Western world is properly quoted as objective evidence of improvement in the general health condition. It is often overlooked, however, that this increase has been due not so much to better health in the adult years of life as to the spectacular decrease in infant mortality. The control of childhood diseases, in turn, resulted more from better nutrition and sanitary practices than from the introduction of new drugs. It is remarkable, in contrast, that little practical progress has been made toward controlling the diseases that were not dealt with by the nineteenth-century reformers. Whereas the Sanitary Revolution did much to eliminate the most common microbial diseases, it has had no counterpart in dealing with the ailments of the adult years of life and of old age.

The nineteenth-century sanitarians believed that health and happiness could be found only through a return to the ways of nature. Modern man, probably no wiser but certainly more conceited, now claims that the royal avenue to the control of disease is through scientific knowledge and medical technology. "Health is purchasable," proclaimed one of the leaders of American medicine. Yet, while the modern American boasts of the scientific management of his body and soul, his expectancy of life past the age of forty-five is hardly greater today than it was several decades ago and is shorter than that of many European people of the present generation. He claims the highest standard of living in the world, but ten per cent of his income must go for medical care and he cannot build hospitals fast enough to accommodate the sick. He is encouraged to believe that money can create drugs for the cure of heart disease, cancer, and mental disease, but he makes no worth-while effort to recognize, let alone correct, the mismanagements of his

everyday life that contribute to the high incidence of these
conditions. He laughs louder than any other people, and
the ubiquitous national smile is advertised *ad nauseam* by
every poster or magazine, artist or politician. But one out
of every four citizens will have to spend at least some
months or years in a mental asylum. One may wonder indeed whether the pretense of superior health is not itself
rapidly becoming a mental aberration. Is it not a delusion
to proclaim the present state of health as the best in the
history of the world, at a time when increasing numbers of
persons in our society depend on drugs and on doctors for
meeting the ordinary problems of everyday life?

Health as Adaptation

Should Virey come back to this world he probably would
experience mixed reactions concerning the success of science in substituting precise knowledge for the lost instincts
of health. He would marvel, of course, at the inexhaustible
spring of new facts uncovered by the scientific method and
at man's skill in converting knowledge into power. He probably would ask himself, on the other hand, whether he and
the Encyclopedists had not taken too much for granted in
assuming that knowledge could be equated with vision and
wisdom. He would see evidence that scientific civilization
threatens to ruin or even to destroy life and creates much
unhappiness whenever it ignores or fails to respect the ethical and emotional values that men prize above life itself.
He would find thoughtful men—untutored persons as well
as sophisticated scholars—fearing that a day may come after
all when "he that increaseth knowledge increaseth sorrow,"
because it is easier for the scientific mind to unleash natural
forces than for the human soul to exercise wisdom and gen-
erosity in the use of power. Even among the most optimistic
he would perceive a disturbing awareness that the solution
of the problems of health and happiness—indeed, their very
formulation—will prove far more complex than had been
anticipated by scientists a few generations ago.

There is no reason to doubt, of course, the ability of the scientific method to solve each of the specific problems of disease by discovering causes and remedial procedures Whether concerned with particular dangers to be overcome or with specific requirements to be satisfied, all the separate problems of human health can and will eventually find their solution. But solving problems of disease is not the same thing as creating health and happiness. This task demands a kind of wisdom and vision which transcends specialized knowledge of remedies and treatments and which apprehends in all their complexities and subtleties the relation between living things and their total environment. Health and happiness are the expression of the manner in which the individual responds and adapts to the challenges that he meets in everyday life. And these challenges are not only those arising from the external world, physical and social, since the most compelling factors of the environment, those most commonly involved in the causation of disease, are the goals that the individual sets for himself, often without regard to biological necessity. Nor can the problem be usefully stated by advocating a return to nature.

It is possible that the haunting memory of the golden age is more than a fond illusion. As suggested by Lewis Mumford, the interglacial periods may have provided a relatively idyllic environment of ease and abundance, breathing spells in the midst of tropical luxuriance that contrasted with the recurrent hardships of the glacial periods contemporary with man's early development. Furthermore, it is also probable that a few people now and then in limited periods of history have enjoyed relative peace in a fairly constant physical and social environment. These periods of relative static equilibrium probably correspond to the era of tranquillity of which the Yellow Emperor spoke, which primitive people often evoke in their legends, and which the philosophers of the Enlightenment had in mind when they pleaded for harmony with the ways of nature. But the state of equilibrium never lasts long and its characteristics are at best elusive, because the word "nature" does not designate a definable and constant entity. With reference to life there is not one *nature*, there are only associations of states and circumstances, varying from place to place and from time to time.

Living things can survive and function effectively only if they adapt themselves to the peculiarities of each individual situation. For some sulphur bacteria, nature is a Mexican spring with extremely acid water at very high temperature; for the reindeer moss, it is a rock surface in the frozen atmosphere of the arctic. Nature for fishes is ocean, lake, or stream, and for the desert rat it is a place where never a drop of water is available. The word "nature" also means very different things to different men. While it is true that human life can occur only within extremely narrow physical limits determined by physiological exigencies, man, by manipulating the external world, renders "natural" for his individual taste many kinds of environments which display an astonishingly wide range of moods. It is a far cry from the equatorial jungles to the Great Barrens, from the Sahara Desert to the fogs of Newfoundland, from the depths of an Arizona canyon to the settlements at 15,000 feet altitude in the Peruvian Andes. Yet man has created a civilization in all these places. As far as life is concerned, there is no such thing as "Nature." There are only homes. Home is that environment to which the individual has become adapted; and almost everything is unnatural outside his range of adaptation.

Harmonious equilibrium with nature is an abstract concept with a Platonic beauty but lacking the flesh and blood of life. It fails, in particular, to convey the creative emergent quality of human existence. Mankind probably achieved identity on the gentle shores of some inland sea with a mild climate, but man in his countless adventures has moved far and wide from the place of his generic birth. It is not with an abstract nature that he has had to deal in his biological and social past, but rather with all the peculiarities that gave a unique character to each of the places where choice, and more often accident, made him stop for a while in the course of his evolution. The seasons and the soil, the plants and the beasts, the permanent dwellers and the distant visi-

tors with which he came into contact during his long jour-
ney, all the factors of his total environment, differed from
one place to another, from one period to another, and their
temporary association constituted the "nature" to which he
had to adapt in each situation and at each moment.

Through the molding forces of biological and social evolution, and by altering the environment to his taste, man
has made himself at home almost everywhere in the world
—in the arctic and in the tropics as well as along the Mediterranean shores—in crowded apartment houses and deso-
lated tundras as well as in the cozy villages of Somerset—in
flimsy tents and in igloos as well as in the châteaux of the
Loire. And it seems to be his chosen fate that he will continue to search for new homes, even though the adaptative
changes made necessary by each move involve unforeseeable dangers. The Garden of Eden, the Promised Land that
each generation imagines anew in its dreams, and all the
Arcadias past and future could be sites of lasting health
and happiness only if mankind were to remain static in a
stable environment. But in the world of reality, places
change and man also changes. Furthermore, his self-
imposed striving for ever-new distant goals makes his fate
even more unpredictable than that of other living things.
For this reason health and happiness cannot be absolute
and permanent values, however careful the social and medical planning. Biological success in all its manifestations is
a measure of fitness, and fitness requires never-ending efforts of adaptation to the total environment, which is ever
changing.

II.

BIOLOGICAL AND
SOCIAL ADAPTATION

The Haunts of Life

Each one of us makes the word "life" mean just what he chooses it to mean. Some use the word to connote a property added to matter but independent of it, a spiritual force that once arose with all its attributes as did Aphrodite out of the foaming sea. For others, life merely represents the summation of effects derived from peculiar arrangements of matter, a set of integrated reactions that became progressively more complex since they began eons ago in the inanimate ooze. The word "life" is also meant to convey experiences, but these are so personal to each individual that they can hardly serve for a characterization applicable to all living things. While it is impossible as yet to define life as an abstract concept, it is fairly easy, on the other hand, to identify many phenomena inherent in *living as a process*. The span from birth to death is occupied by countless reactions to the environment which allow the organism to build and maintain itself. And the responses that the organism makes to the stimuli originating from the environment constitute the inward and outward manifestations of its existence.

Whatever the origin of life, its maintenance demands that living things achieve some sort of fitness to the environment in which they develop and function. Through the processes of evolution life continues to emerge, fitting itself to all corners of the earth and endlessly changing as the environment changes. This marvelous plasticity, the ability of living things to develop structures and functions that permit them to solve problems imposed by local peculiarities, has been a source of everlasting wonder. Turn up a stone, dig a hole, explore a cavern, dive under water, or climb on a rock; each situation reveals plants and animals adapted to it and hardly ever found anywhere else. The naturalist—this eternal child—never ceases to marvel at the peculiarities of each haunt of life. Anton van Leeuwenhoek, observing for the first time in 1673 the microscopic inhabitants of his mouth and of the mud in the Dutch canals, experienced the same fascination that the clergyman Gilbert White and the school teacher Henri Fabre experienced one and two centuries later in studying, respectively, the swallows in Shelbourne and the insects in Provence.

New examples of nature's resourcefulness continue to turn up. In industrial areas of England, or in the districts where prevailing winds cover the trees with black soot, dark forms of certain species of moth tend to predominate; the pale-colored forms, in contrast, gain the upper hand in unpolluted districts. It seems that birds act in each case as one of the selective forces, detecting more readily and thus destroying more effectively the phenotypes of moth with a low camouflage efficacy—the pale insects against the smoked bark or the dark mutants against the lichen of the tree trunks. True enough, most modern biologists pretend to be concerned with more subtle and more sophisticated problems and are prone to take a somewhat blasé attitude toward the findings of old-fashioned naturalists. But, while many generations of scientific discovery seem to have dulled the primeval appetite for wonder, there persists the urge to understand the processes through which life manages to make into its own the resources of the inanimate world. Scratch the surface, furthermore, and the sense of wonder often comes to light even in the most matter-of-fact scientists. It is not only to discover physicochemical explanations that men become engrossed in the radar-like mechanism which guides the flight of the night owl; in the resistance of the deep-sea animals to very high pressure; in the blindness of the creatures which thrive in places where light never penetrates; in the ability of the desert rat to live without water. A rational explanation can be found for any one of these extraordinary feats of nature. But it remains akin to a miracle that every spot on earth has become a haunt of life, harboring several types of living things so well adapted to it as to crowd it to the fullest of their biological potentialities.

The only way to reach a complete understanding of the mechanisms by which matter can become converted into so many forms of life would be to watch the evolution of living things from their origin and as they progressively adapt themselves from one type of environment to another. For, as Aristotle wrote in his *Politics*, "we shall not obtain the best insight into things until we actually see them growing from the beginning," thus rephrasing Heraclitus of Ephesus' statement that "he who watches a thing grow has the best view of it."

If life were measured only in terms of physical characteristics and biochemical activities, man could not claim any trait differentiating him fundamentally from the rest of the creation, let alone placing him at its top. Men and microbes utilize sugar and other foodstuffs in much the same way and their genetic equipment is very similar in structure and in function. The biochemical unity of life and the similarity in mechanisms that it uses for the transmission of hereditary characteristics are now accepted biological laws. Indeed, there are many biological machines with more ingenious parts than man can boast of and possessing greater efficiency and robustness. But if life is regarded as emergence, displaying at each step in its evolution new properties that do not appear necessarily implied in the past and therefore could not be predicted from the constituent parts, then man can claim to be its most spectacular achievement. For he has moved furthest from the limitations of the protoplasmic ancestry that he shares with all other living things. More important, he continues to move away from it at an ever-increasing rate and even pretends now, probably for the first time in biological history, that he can direct the course of his evolutionary march.

While all living things depend upon the same fundamental metabolic processes, the various haunts of life display much biological specialization in detail. Most organisms are highly selective in their requirements and must in
addition compete with organisms of their own and of other
species for available resources. As a result, they are severely
restricted geographically with regard to the conditions that
are optimum for their existence. In contrast, man, despite
his highly developed biological personality, has succeeded
in colonizing a large part of the surface of the earth and is
now contemplating the invasion of outer space. The persistence and ingenuity displayed by the human species in moving ever onward from its geographical origin could hardly
have been predicted from its physicochemical make-up. It
is only by developing processes of social adaptation that
mankind has been able to achieve such extraordinary biological success. Needless to say, these social processes were
superimposed on the results of adaptive mechanisms similar
to those which operate in the rest of the living world. It
is therefore as an animal that we must first consider man
in his struggles with the environment. For man evolved as
an animal, even while he was dreaming of God and the
stars.

Biological Adaptations in Man

While all members of the human species apparently have
common origins, they constitute today several obviously dis-
tinctive biological subgroups, and it is apparent that each
is particularly fitted to certain environmental conditions.
These conditions were certainly those prevailing in the lo-
calities where the subgroup remained semi-isolated for long
periods of time, either out of choice or, more frequently,
as a result of accident. For there is no doubt that, just like
other living things, man was molded and chiseled into shape
by his physical environment. During countless generations
he remained the slavish, helpless creature of this environment, and had to submit to its limitations in order to survive

and proliferate. It is true that the endless mingling of peoples that resulted from mass migrations usually makes it impossible to recognize with certainty in any given type the reflection of the physical past. Nevertheless, a few characters seem to possess such obvious selective value that they can serve to illustrate the adaptive processes through which mankind evolved.

From classical Greece to the Renaissance certain proportions of the human body have been accepted as approximating perfection, both for biological effectiveness and for physical beauty. Granted a few aberrant tastes developed for the sake of diversity, the Greco-Roman world and its descendant, the Western civilization, have agreed on a formula of ideal body frame. This ideal was probably suited to the climatic conditions of the Mediterranean shores, conditions that Western man still regards as optimum for his comfort and tries to create artificially if nature does not provide them where he happens to sojourn. But men have also evolved in other climates. In many parts of the world the average body proportions, as well as the ideal of physical beauty expressed in plastic arts, reflect physiological exigencies very different from those of life on the Mediterranean shores. There is no doubt, for example, that the prevailing temperature and humidity bear some relation to the shape of the body, since the smaller the ratio of body surface to body volume the smaller the heat loss per unit of time. A short, stocky body frame covered with fat helps the Eskimo economize body heat in the arctic climate. In contrast, some tribes of Equatorial Africa exhibit a tall, lanky, gracile structure which probably helps in dispelling body heat. It has been suggested also that tallness is of advantage in the desert because the temperature in dry countries is higher at the ground level than at a few feet above.

Needless to say, many factors other than heat and cold have influenced the evolution of body shape. On level lands where it is often helpful to see far and move fast, tall men are at an advantage over short-legged, rotund ones, whereas the situation is different in densely wooded areas The forest dweller must climb over logs, pull obstacles from the ground, push aside branches and vines, twist his body in order to proceed. Thus displacement in the forest depends upon the use of a variety of muscles that are less often called into play for locomotion on the plains. As a result, the forest dweller is likely to be short legged, long trunked, barrel chested, and broad handed. A stocky muscular build is as important for his survival as the greyhound body form is of advantage to the plainsman.[1]

Evolutionary history is often reflected also in the distribution of muscles and the structure of bones. People who live in forests or are compelled to plod through snow are more likely to be at an advantage if they possess heavy bones, whereas the swift runner is better served by long light bones. Above and beyond these obvious morphological characters, there are many physiological adaptations of a more subtle nature which can be related to the physical environment. The greater the amount of blood per body weight, for example, the more efficiently the sweating mechanism permits cooling in hot dry climates.

The color of the skin also suggests obvious adaptation to the physical environment. Certain skin pigments can shield the body from ultraviolet light and from other forms of radiant energy which are harmful if excessive. On the other hand, penetration of some ultraviolet light into the skin is essential for the synthesis of vitamin D in the body and may have other beneficial effects. On the basis of these simple facts it is easy to rationalize the prevalence of darkness of the skin in some parts of Africa and of pale skin colors in regions where cloudy skies prevail, as in Northwestern Europe. It is true that the exceptions to the rules linking body build and skin pigment to climate and radiation are so numerous that the adaptive value of these physical characteristics may seem problematical under the natural conditions prevailing today. But they were probably of much clearer relevance during the early phases of man's biological past.

[1] See for example Carleton S. Coon, *The Story of Man, From the First Human to Primitive Culture and Beyond*, Alfred A. Knopf, New York, 1955, 437 pp.

The study of small human groups living under unusual
environments often reveals strange adaptive mechanisms.
The aborigines in Central Australia, for example, face
extraordinary problems of water supply.[2] From one hunting
ground to another they may have to make treks of 100
miles or more across desertic land where the rainfall is less
than 10 inches a year and the temperature may reach
140° F. In addition to having developed an extraordinary
instinct for discovering water where white men would die
of thirst, the aborigines exhibit physiological adaptations
that permit them to survive with very small amounts of the
precious fluid. They are able, for example, to use their
stomachs as water bottles in which large volumes can be
stored. Their enormously distended stomachs are the evidence of storage whenever they start for a trip across the
desert from a place where water is available. A European,
drinking large quantities of water, rapidly excretes the
excess once his physiological requirements have been met.
In contrast, the stomach in the aborigines is able to retain
the water and to let it out as needed, over many hours.
Furthermore, their kidneys seem to be so efficient that they
apparently require only half as much water to flush the same
amount of waste products as would be the case for white
men—thus reducing greatly their minimal requirements.

One of the strangest adaptations to water supply has been
recently observed in a large town of North Africa. The only
ground water used by the inhabitants of this town is extremely high in sodium chloride (3,000 parts per million).
This water is unacceptable to outsiders, but the inhabitants
are so accustomed to it that they take salt with them on
their travels to mix with their coffee! In contrast, there have
been and still are many groups of men all over the world
who do not consume salt regardless of the type of food
eaten. Indeed, some tribes in Africa use wood ashes, rich in
potassium, as a condiment instead of salt.

The Aymara Indians of the Lake Titicaca region in the

[2] This information was kindly given me by Dr Charles Lack
of the Royal National Orthopaedic Hospital, Brockley Hill, Stan-
more, Middlesex, England.

Peruvian Andes illustrate the ability of man to adapt to life at very high altitudes. The large chests of these Indians, their great depth of respiration, and the richness of their blood in hemoglobin permit them to engage in vigorous physical exertion in areas above 12,500 feet, where unadapted men of other races are handicapped by the low oxygen supply.

In addition to hereditary adaptive mechanisms which are so obvious as to be discernible by the naked eye, there are many others so subtle that they are revealed only by accidental observations or by systematic laboratory studies. One example, so strange as to appear irrational, is the relation recently discovered between the sickle-cell trait and resistance to malaria.

It is known that a large number of Negroes—up to forty per cent of the total population in certain tribes of Africa —carry the sickle-cell gene which is associated with a certain form of anemia; four per cent of these natives receive the gene from both parents and hence are subject to the disease, sickle-cell anemia. A child inheriting two sickle-cell genes has only about one fifth as much chance as other children of surviving to reproductive age, with the result that about sixteen per cent of the sickle-cell genes must be removed in every generation. Yet there is no sign that the level of incidence of these genes has decreased in Africa during historical times. This surprising fact makes it necessary to assume that the sickle-cell trait gives some survival advantage to individuals who carry it over those who do not, thus counterbalancing the higher mortality of sufferers from sickle-cell anemia.

Recent discoveries have provided a spectacular and most unexpected confirmation of this hypothesis. In brief, it appears that children with the sickle-cell trait exhibit an unusually high degree of resistance to malaria. As exposure to this infection so far has been almost constant in most of Central Africa, the resistance conferred upon children by the sickle-cell trait permits them to develop during youth a resistance to malaria that persists throughout the rest of their lives. Consistent with this interpretation are the facts that in Africa the sickle-cell trait is frequent only among the Negroes living in malarious areas, and that it occurs also among non-Negroes wherever malaria is prevalent, for example, in southern Italy, Sicily, Greece, Turkey, and India.

If the persistence of the sickle-cell trait is associated with resistance to malaria, its frequency should fall whenever a population moves from a malarious area to one free of this infection, since the trait would no longer provide a survival advantage under these circumstances. Interestingly enough, this is apparently what is happening in North America, where the incidence of the trait among Negroes has now fallen much below that prevailing in the parts of Africa from which the slaves were introduced into the New World. This phenomenon provides an example of change in genetic structure, detectable within two centuries after the change in environment, in this case determined by the transfer of the Negro from Africa to North America.

Many adaptive mechanisms are not permanent characters of the adapted individual but rather depend on temporary responses to the environment. These transient changes permit each individual to accommodate to new situations more rapidly than can be done through the agency of the hereditary processes just discussed. The ability to develop tan, for example, can afford within a few days a somewhat increased resistance to ultraviolet radiation. Similarly, some degree of transient adaptation to life at higher altitudes can be rapidly achieved by changes in the mechanics of breathing, in the blood hemoglobin, and in tissue metabolism. In many cases the changes are reversible, the individual returning to his previous status when removed from the environment that had elicited the adaptive response.

Among nonhereditary responses should be mentioned all the mechanisms which increase the resistance of the individual to the poisons and microbes ubiquitous in his environment and with which he therefore constantly comes into contact. It is just as well that words like "immunity" and "tolerance," which are used by scientists to connote resistance to infection, are still poorly defined, for they will have to accommodate many unexpected discoveries in the future.

For example, a new fact was brought to light very recently by the use of animals raised in an environment entirely free of bacteria and molds. These so-called germ-free animals appear essentially normal and are able to reproduce themselves for several consecutive generations in the man-made, germ-free world. But they are highly susceptible to the common ubiquitous microbes, even to those usually unable to cause disease in animals raised under less esoteric conditions. The germ-free animals commonly develop severe infections, of which they die, as soon as they are placed in the open world. It is of interest in this regard that their internal organs are very deficient in certain constituents always present in normal animals, in particular the cells making up the so-called lymphoid tissue, which presumably plays a part in resistance to infection. This tissue appears, therefore, to be produced normally as an adaptive, protective response to the microorganisms which contaminate all ordinary objects and to which all living things are exposed constantly.

Thus, life in the world of nature, implying as it does endless contact with all kinds of microbes, early brings forth in animals an adaptive response which modifies the internal organs in such a manner as to increase their general resistance to infection. Recent investigations strongly suggest that the same holds true for newborn babies. It was easy to imagine that life in an aseptic world might be rather dull; it appears now that it might also be dangerous in the long run, by rendering the too-well-isolated individual unable to survive in any world except that in which it has been raised.

Adaptation through Instinctive Patterns of Behavior

Acquired immunity, whatever its mechanism, is a sort of biological learning which makes the body better prepared to meet a danger already encountered in the past. The spectrum of changes that can occur in an organism as a response to prior experience is very broad. Biological learning can range all the way from the production of antibodies detectable by chemical techniques in the serum of an immunized individual, to the wisdom of the crafty old man experienced in the ways of the world, and to the social knowledge which schools and books try to convey by describing the various aspects of human experiences. There is an unbroken continuum from the wisdom of the body to the wisdom of the mind, from the wisdom of the individual to the wisdom of the race

Many examples could be quoted to illustrate that learning can occur at all levels of biological organization. Even a lowly protozoan can learn and remember from prior experience that contact with a solution which is not injurious in itself may foretell of a more dangerous experience to follow. The experienced protozoan can take advantage of this awareness to escape before the time of danger has arrived. So developed is the faculty for learning even in the lowest organisms that, as we shall see later, it makes them respond as violently to symbols of danger as to the danger itself. Would that all human beings might acquire thus readily from the first cocktail a warning of the dangers present in the third!

Adaptive patterns of behavior become, of course, increasingly important as one ascends the scale of living things. In higher animals, and more so in man, adaptation expresses itself in instincts, tastes, and habits which help the group and the individual in making use of available resources and in avoiding sources of danger. Thus, wild animals display little fear of travelers where man is rarely seen, for example, in the antarctic or in many national parks and game reserves where they are protected. In general, they seek to avoid man only when their past collective experience makes them consider him a potential enemy. Animals, like men, have a subconscious social memory.

Any population long settled in a fairly constant environment develops likes and dislikes, which are the outcome of countless trials and errors. From this experience emerge a variety of customs, taboos, religious beliefs and practices which permit that population to survive under conditions that are often unacceptable or fatal to inexperienced newcomers. Among the countless examples that could be selected to illustrate this subconscious social wisdom, nutritional habits are particularly illustrative because their incredible diversity often accounts for the survival of many primitive people under conditions that appear at first sight incompatible with human life. Furthermore, a rational basis, meaningful in physiological and chemical terms, has been commonly found to justify local food habits developed empirically.

Automobile transportation, laborsaving devices, and overheated dwellings have so greatly reduced the expenditure of physical energy in much of the Western world that make-believe foods low in calories now have a premium value. But physiological exigencies were very different in the past, and still are in many parts of the world. In the arctic, food had to provide man with the ability to resist cold as well as to exert great physical effort, and the Eskimo solved this problem by consuming a diet extremely rich in fat. Many primitive peoples use a wider range of food than civilized peoples, and after living for many generations in a stable environment they usually develop through trial and error an empirical knowledge which permits them to derive adequate nutrition from apparently limited resources. Despite the scarcity of meat and milk in ancient China, for example, each local community had worked out a diet with an amino acid composition of adequate nutritional value through the skillful mixture of plant products readily available in that particular region. Insects, mushrooms, teas from evergreen needles, berries, and other wild fruits have constituted sources of proteins and vitamins for countless generations among men who never thought that they were thus protecting themselves from deficiency diseases.

Many primitive practices of cookery also appear to have some rational basis. The Mexican peon learned to grind corn in chalk and thus empirically enriched his diet with the calcium in which it would have been otherwise deficient. Pulque, the fermented juice of the agave plant, is for him not only a pleasant alcoholic beverage; the suspended matter which gives pulque its milky appearance provides in addition an abundant source of various vitamins. All ancient techniques for the processing of milk into products ranging from the various forms of kumiss or yoghurt in the Near East to the most sophisticated cheeses of Western Europe served a double purpose Ripening increases the digestibility of casein and lactic acid fermentation interferes with the multiplication of disease-producing bacteria in the milk. It remains to be proved that pasteurized milk distributed in a paraffined container has any advantage from these points of view over well-prepared Camembert or Emmentaler cheese. Biological wisdom appears even in the ritual of the cocktail party with its hors d'oeuvres rich in fat which retard the absorption of alcohol and thus minimize the stunning effect of the preprandial drink.

Adaptation through Social Mechanisms

The biological mechanisms of adaptation which have physical and physiological bases, as well as those resulting from instinctive patterns of behavior, are supplemented in man by conscious social processes. In fact, it is certain that social adaptations have been the most influential determinants of man's fate during historical times. Human life is now molded to a large extent by the changes that man has brought about in his external environment and by his attempts at controlling body and soul.

Since he began to use shelter, clothing, and fire man has gone far in his attempts to achieve fitness through control of the external environment. By techniques which permit him to produce and transport at will not only food and drink but also warmth or cold, dryness or humidity, Western man has been able to surround himself wherever he goes with an atmosphere of his choice, probably similar to that in which he has biologically evolved. A few cubic feet of Mediterranean-like climate are sufficient to give him the illusion that he can colonize any part of the earth, be it the tropical rain forest or the frozen poles. But in many of the places that he occupies he has come as a conqueror, living with his imported ways instead of adapting himself to become a true habitant. In so doing he has adopted a policy that places him outside the natural order of things. By changing the physical world to fit his requirements—or wishes—he has almost done away with the need for biological adaptation on his part. He has thus established a biological precedent and is tempting fate, for biological fitness achieved through evolutionary adaptation has been so far the most dependable touchstone of permanent success in the living world.

To some extent, it is true, man has also tried to control and even to modify his own physical and mental self. His internal environment is not so rigidly fixed as was once assumed and it is now apparent that slight modifications of it, still compatible with life, can exert an enormous influence in determining capabilities, behavior, urges, even goals. There is no doubt that the whole pattern of life, for an individual or a group, can be profoundly altered by the type and amount of food consumed and by microbial activities. Furthermore, it is now possible with the proper hormones or drugs to manipulate almost at will the physical appearance and the behavior of any given person. Synthetic stilbestrol is utilized today for the chemical caponization of roosters and for driving chickens to more efficient egg laying. Someday, no doubt, love philters will be sold over the counters, drugs more dependable than whisky will help timid men to behave for a while like heroes, and other drugs may be used to direct life toward goals that will determine the future trend of human evolution. In reality, this will not be entirely new. Insects discovered millions of years ago that the feeding of royal jelly to a larva could transform it into a queen bee unwillingly committed to endless egg laying. Much more recently, but still several centuries back, a few family groups in Scandinavia had learned the secret that the eating of certain mushrooms made them oblivious to pain and to danger. Under the influence of the holy mushroom they fought unprotected and terrorized their opponents. If tradition does not mislead us, it is from their habit of going in bearskin that they gained the name "Berserk" and that the expression "going berserk" came to designate certain deviations from normalcy. Empirically, demagogues and dictators have long known how to distort public opinion according to their will. Advertising has now enlisted the participation of behavioristic science and it is apparent that the mind can be molded into almost any shape with even greater ease than the body.

The most challenging prospects now come from the possibility of affecting the very hereditary make-up of man. By mating the proper pairs of plants or animals, skilled breeders have known from time immemorial how to produce new varieties of corn, dahlias, pigs, or dogs for the benefit or the amusement of man. The potentialities of genetic control bid fair to outdo these empirical achievements, since it is now possible, at least in bacteria, to change the hereditary make-up without the need of sexual union. But, without going so far, it is sufficient to remember that artificial insemination is a widespread practice in animal husbandry and that it also works in the human species. Who can doubt that human eugenics will eventually overcome the resistance of social traditions and ethical scruples and will make it possible to plan parenthood not only in time and quantity, but also in quality. A day may come when children can be made almost to order, with perfect fitness for life in the arctic or on the equator, in the foundry or in the presidential chair.

Unstability of Ecological Systems

Living organisms cannot be defined or even described merely in terms of their activities or of the constituents of their bodies. Under any conditions that support life its creatures tend to organize themselves into highly integrated systems which involve all the forces of the environment, lifeless and living, atmospheric and geological. As we have seen, there is evident everywhere in nature a close correspondence between most of the structures and activities of living things and the needs imposed upon them by their places in nature. The necessary outcome of this relationship is a dynamic equilibrium which implies inseparability of life and environment, and which has fitness as a consequence. It will be helpful to illustrate with a few examples the extent to which all ecological systems are susceptible to disturbances originating from without or from within.

Nothing at first sight could have appeared more innocuous than the introduction of a small animal into an island where it had not existed before; yet several such events have had large and unexpected consequences during recent times. The mongoose, a small rodent native to India, was introduced in 1872 into the West Indies island of Jamaica in an attempt to control the rats which caused great damage in the sugar-cane fields. Although the numbers introduced were small—four males and five females—the mongooses multiplied so fast that they destroyed most of the rats within ten years. Unfortunately, they then turned to other small creatures and have now all but exterminated much of the native fauna—mammals, reptiles, and ground-nesting birds, thus becoming pests worse than the rats. In 1931 nine ermines were imported into the Netherlands to control rats and rabbits. Within twenty years they had themselves become such a menace, preying on wild birds and poultry, that the Dutch government had to put a bounty on them. Early in the 1940's a local zoo on the Japanese island of Oshima exhibited a few squirrels from Formosa. The squirrels soon overran the whole island and almost ruined its major industry, the production of camellia oil, by destroying the birds that distributed the pollen to the camellia flowers. On a less dramatic scale, the English starling is becoming a menace for some of American birdlife. Introduced from Europe on the Atlantic coast (apparently in Central Park, New York, in 1890), it has been spreading across the continent ever since, progressively displacing other, more attractive birds. The most lovely harbinger of spring, the bluebird, is one of the species least able to maintain itself in areas where the starling becomes established. As we shall describe in the following chapter, the introduction of the rabbit into Australia illustrates still another aspect, even more spectacular, of the unexpected results that can follow any disturbance in ecological systems.

Tremendous upheavals due to the introduction of a new organism into an ecological system have often occurred without the conscious participation of man. New plant pests, in particular, have been imported unintentionally from one continent, in which they were essentially innocuous and had remained unnoticed, into another, where they proved immensely destructive to one or several plant species. This is the case with the phylloxera, a small aphis normally present on the American grapevine, to which it causes little damage. Around 1859 the insect was introduced into France on American vines. From there it rapidly spread through all the grape-producing sections of Europe, killing the vines everywhere. In fact, the susceptible varieties of European grapevines appeared doomed to destruction when it was fortunately discovered that they could be saved for production if grafted on the roots of American vines resistant to the pest. It was due to this fortunate circumstance that the European grape and wine industry could survive—the American roots provided the vigor and the European graft the quality. Similarly, the rhinoceros beetle introduced into some of the Polynesian islands during World War II is threatening to destroy the coconut tree, one of the most valuable crops of these areas.

For the American woodland the chestnut blight has had effects as dramatic as those of the phylloxera in France and the rhinoceros beetle in Polynesia. The chestnut was admirably adapted to conditions prevailing on the American continent, as witnessed the immense stands of stately trees still common a few decades ago. Yet, in the fifty years since the fungus responsible for the blight was first discovered in the United States, it has practically killed the American chestnut throughout its native range. The infective spores of the fungus are distributed over short distances by wind and water and over longer distances by woodpeckers and sapsuckers whose beaks become contaminated while feeding on insects in cankers of infected chestnuts. As is well known, this handsome and useful tree has been almost completely eliminated from large areas where it once flourished.

More frequent and theoretically more interesting than these dramatic upheavals caused by the introduction of foreign pests are the upsets of ecological equilibrium resulting from disturbances of internal origin. We shall describe in Chapter IV the far-reaching effects exerted by changes in weather conditions on the fate of the potato crop in Ireland and on the destiny of the Irish people.

Social Determinants of Human Fitness

Man, needless to say, is as susceptible as are animals or plants to sudden changes in his physical environment and to attack by new parasitic invaders. No technical knowledge can ever protect him completely from assaults and disturbances that he cannot possibly foresee. As many examples of unexpected external threats that changed the course of history will be presented in subsequent chapters, it does not seem necessary to labor the point further at this time. Instead, we shall consider now another type of disturbing influence which is apparently peculiar to mankind and has its origin in the fact that physiological requirements are no longer the mainsprings of human behavior. There is no doubt that man's own caprices and vagaries constitute the most insuperable obstacles to the achievement of the millennium and to the success of utopias.

It is a universal trait among men that as soon as their physiological needs are satisfied they develop new wishes and urges, which in turn are soon replaced by other desires. In man, at least, satisfaction is commonly followed by boredom. Food in abundance, supplying all the required nutritional elements, may solve the problem of eating for animals, but it does not do so for man. Even the most pampered cat or dog will happily eat day in and day out a standardized diet canned in Chicago with scientific but monotonous perfection. In contrast, the more man becomes civilized or at least urbanized, the more he is likely to lose

the experience of honest physiological hunger and to replace
it by nonphysiological needs born out of the pleasure of
eating. This pleasure soon becomes an end in itself, replacing the physiological purpose from which it had originated.
Cabbage and turnips would provide all the ascorbic acid
required by the human body, yet millions of American
mothers would believe their families on the brink of vitamin starvation unless fresh orange juice was on the breakfast table every morning. Two centuries ago tea and coffee
were at best sophisticated luxuries for the favored few. The
American citizen today believes that he cannot function
effectively if deprived of caffeine stimulants. The taste and
color of the dietary product, the package in which it is
distributed, and the customs associated with its use are as
essential factors in human nutrition as are the intrinsic values
of the food or of the stimulant. And it is well known that
fashions in tastes, colors, packages, and customs change fast
in the modern world. The shelves of the supermarket constitute the colorful and rapidly moving record of the trans-
formation into life necessities of tastes and habits newly
acquired from social or advertising pressures.

In brief, it is certain that social criteria are far more influential in the selection of food than are preoccupations
with optimal growth rates and with the adequate performance of biochemical processes. In order to be acceptable,
diets must of course satisfy the needs of the body, but even
more imperatively must they be compatible with the habits
of the group as well as with the respect of its taboos—and
this often imposes strange strictures upon the chemical concepts of nutrition. The changes in ideals of women's figures
in the course of ages reflect well the capricious basis of
dietary customs.

While scientists can cite the Venus of Milo and a few
Greek-inspired goddesses as famous examples of feminine
beauty compatible with scientific nutritional standards, it
seems that this physical type is more popular as a museum
piece than in the world of the living. In some parts of Latin
America and of the Orient the plump, well-padded woman
has far greater sex appeal, hence the sweet desserts which

supplement her meals and occupy her idle hours. With us, in contrast, it is the tall, lanky girl who becomes the fashionable model or actress; hence her vitamin-rich but calorieless breakfast and the leaf of lettuce for lunch. The desire to look emaciated is not new in the history of fashion. In the sixteenth century Montaigne wrote of women swallowing sand to ruin their stomachs in order to acquire a pale complexion. During the nineteenth century the proper attitude at a romantic dinner was affected indifference to food, and it became fashionable among men to pretend a passion for languid, ethereal womanhood. As a response, young women were inspired to drink lemon juice and vinegar in order to kill their appetites and thus become more attractive. Repetitious indeed are the tricks used by the sexes to attract each other.

Loss of biological wisdom in food habits does not occur only as a result of distorted Western civilization. It is just as frequent in other parts of the world and even among animals. There is no such thing as an instinct of good nutrition. There is only the kind of empirical learning that comes from trial and error, from experiences gained subconsciously under a given set of conditions. But this kind of subconscious nutritional wisdom is lost as soon as conditions change, in particular if they change too rapidly. A recent study of malnutrition among South African Zulus revealed the havoc wrought by the substitution of millet for corn in their diet. Millet had been the indigenous cereal before the arrival of the white people, but was displaced by corn when the natives were shown that the latter crop yielded more abundant food for less work. Other changes occurred almost simultaneously. In the past the large areas of unsettled land had provided the Zulus with game, milk, and berries, products which eventually disappeared from their fare as South Africa became colonized. Yet the changes in diet were so progressive that they remained almost unnoticed by the Zulus themselves. Upon inquiry, they stated with conviction that their present diet was the same as that which had made their ancestors strong and healthy and they could not believe that their disease problems of today were largely the result of change in food habits.[3]

Animals have no better natural instinct of nutrition than human beings when placed under strange circumstances. The proper diet of a wild animal recently captured can be discovered by presenting the animal with several alternative foods and noting the ones that it prefers. But this technique gives misleading answers when applied to animals accustomed to captivity. Inmates of zoos frequently consume by preference inadequate foods which have become palatable to them through habit, even though it causes nutritional disturbances.

Advanced as it is, nutritional science can compensate for only part of the lost knowledge of adequate nutrition formerly gained through experience, because it is concerned with only a limited aspect of human life. Even when all the information necessary for optimum growth of chickens, rats, and men is available, there will still remain to be considered many important effects of food products other than those determining ability to support growth. It was discovered a few years ago, for example, that several pasture plants, in particular subterranean clover, contain significant amounts of substances endowed with estrogen activity similar to that exhibited by some of the sex hormones produced in the body. These estrogenlike substances appear to be most abundant in the young new growth during the early part of the year. For the farmer, the existence of varying amounts of sex-hormone-like substances in herbage is of importance because of their effect on fertility, lactation, growth, and fattening in grazing animals. But human behavior also is under endocrine control, and one may wonder whether the consumption of tender sprouts and young leaves at the time of nature's awakening does not contribute to making of early May a biological as well as an emotional experience. There is much more to food than the factors

[3] Quoted from John Cassel, "A Comprehensive Health Program among South African Zulus," in *Health, Culture and Community* (ed. Benjamin D. Paul). New York: Russell Sage Foundation, 1955.

known to affect the growth rate of laboratory animals and even of human babies.

Because it affects the daily life of every person in a fairly direct and obvious manner, nutrition constitutes a convenient subject to illustrate the vagaries which render so unpredictable the future trends of social evolution. But any other aspect of life would provide as telling examples. Thus, excessive concern for security and for comfort may indirectly cause hereditary changes that decrease man's ability to meet biological threats. The wild Norway rat is fierce, almost untamable, but has an extraordinary ability to resist all sorts of stresses. In contrast, its laboratory-bred counterpart is docile and easily handled but reacts poorly to stress and fatigue, probably as a result of atrophic changes in its adrenal glands. Well-domesticated man, although highly desirable as a good citizen, may prove to be a poor biological specimen when placed outside the sheltered social environment to which he has learned to conform.

The comparative value that society puts on the athlete or the aesthete, on the family farm or on the anonymous assembly line, is not without influence on the type of fitness regarded as desirable. Moreover, social planning, conscious or unconscious, does not necessarily result in the desired effect. If the first creature seen by a newly hatched gosling is a human being, the bird will accept this person as a mother object and readily follow the lead of its honorary parent. Like geese, men also may be early influenced by odd circumstances and these can hardly be ruled out or introduced at will even in the most completely organized social system. Social evolution may be the result of intention, but it rarely, if ever, produces the result intended.

The desire to imitate a fashionable type can also give peculiar orientations to the development of human traits. Literary and iconographic sources suggest, for example, that the presence of a goiter was considered a highly attractive attribute of women in some parts of Europe during the sixteenth century, and the trait can often be recognized in provincial religious iconography obviously intended to express loveliness and to exert emotional appeal. There is some evidence that idealization of this pathological type resulted in an increased prevalence of goiter in certain areas, perhaps through sexual selection.

It is true, of course, that fashions have usually been too evanescent to exert a significant influence on the hereditary make-up of human groups. The contrast between the male robustness of Republican Rome and the delicacy of the Florentine Renaissance probably corresponded to a phenotypic manifestation rather than to genetic differences. And, despite the testimony of countless entertaining photographs, the flapper type of the 1920's and the Hottentot woman with her pendulous breasts may be less apart in genetic endowment than would appear from the way they look ideally to their respective menfolk or to the selective eye of the reporter's camera. Nevertheless, these contrasting types serve as evidence of the fact that bodily appearance and emotional type are the products not only of the physical environment but even more of the ideas evolved in the mind of man.

Whatever the mechanisms of their effects, genetic or environmental, the vagaries in human tastes and goals will certainly continue to produce in the future, as they have in the past, unexpected and peculiar kinks in the evolutionary line. This may be regrettable for the biological destiny of man, considered merely as a living machine, but it has contributed much to enlarging the range and interest of history.

Comparing human behavior and attitudes in the course of a trip from Paris, through Flanders, the Netherlands, Germany, Alsace, back into the Champagne country and the Ile de France, Hippolyte Taine became convinced that the genius of each country was the direct expression of the characteristics of its land, the color of its skies, and the moods of its climate. He might have reached a different philosophy had he traveled also through the American Southwest. There he would have seen two very different types of culture coexisting side by side—the Hopis, who live a sedentary, agricultural existence in crowded adobe settlements, carefully husbanding the scarce water to raise a few crops, and the Navajos, who move periodically from one isolated family hogan to another, with a pastoral type of culture based on sheep and goats. In addition to the Hopis and the Navajos, many different Indian tribes have achieved remarkable fitness to the exacting conditions of the American Southwest—as have also the Mormons, the Catholic Mexicans, and godless men in search of adventure. The fitness achieved by these diverse human groups was not merely the result of passive adaptive processes such as those which produced the cactus and the desert rat. It was a creative process in which man selected conditions and exploited local forces to impose on nature a pattern of his own choice.

It is certain that in order to survive man will need to achieve many new adaptations to the effects of the very forces that he himself is setting in motion. The kind of fitness that made a good soldier in Attila's bands, in the Greek phalanx, or in Napoleon's armies might not be appropriate for the handling of jet planes in future warfare. The men best adapted to the electronic age are likely to differ physically and emotionally from those who cleared the Ohio forests. Yet the foot soldier will retain some essential part in modern warfare and there will always be need for the hewers of wood and drawers of water even in the automation age. Indeed, human societies may come to encourage or to compel the development of different kinds of biological specialization among individuals. This has happened among social insects in which individuals exhibit regression or even disappearance of certain organs and functions to permit the extreme development of other organs with a high degree of specialized activity. Colonies of ants have queens and kings whose only business is reproduction; sexually underdeveloped females which function either as workers to collect food and build the nest or as soldiers to fight intruders; forms with peculiar heads shaped to plug the nest entrance and others with enormous abdomens to store food. As shown by the success of the social insects—ants, bees, and termites—this type of society may remain stable, vigorous, and viable during long periods of time. Nevertheless, it seems incompatible with the ideals which have given its uniqueness to human life.

Even if it is not necessarily true that there exists a reciprocal relationship between breadth of specialization and likelihood of racial extinction, as suggested by evolutionary records, man must continue to evolve in order to remain true to his racial genius. For this reason the more civilization increases in complexity and the more it compels its members to become specialized the more it is necessary to maintain a certain number of human activities in a primitive, unorganized state. In a wise society leisure and holidays—instead of becoming stereotyped as they presently are —should play a role similar to that of national parks and wildlife reservations, where plants and animals retain some chance to practice the mechanisms which have permitted evolutionary adaptation.

Fortunately for his biological and social future, man has in reserve a large store of unspecialized tissue cells in his body and many unexploited potentialities in his brain that still permit him to evolve adaptively and achieve fitness to many unexpected situations. But fitness will never be a static condition with which he can be satisfied. All living things and their environments change endlessly and no permanent equilibrium between them can ever be reached. Man has added still further complexities to the biological situation by urges and strivings that have nothing to do with species survival. As long as he continues to reach further into the unknown and to create for himself situations governed by parabiological values, problems of adaptation will endlessly arise. For him, certainly, fitness is more than a biological end; it is a state constantly to be modified and even sacrificed for the sake of new illusions and new goals.

III.

STRUGGLE AND PARTNERSHIP IN THE LIVING WORLD

The Struggle for Existence

Ironically enough, it was in the writings of two of the most gentle heroes of biology—Charles Darwin and A. R. Wallace —that the Victorian era found the scientific authority for its blood-curdling slogans. To Spencer and his contemporaries the theory of organic evolution meant above all, if not exclusively, that the struggle for existence assured the "survival of the fittest." The Victorians were pleased to count themselves among the survivors in this natural weeding out of the unfit. In their view the law of the jungle was also the law of social progress, and gave scientific sanction to laissez-faire economics, imperialistic policies, sweatshops, child labor, and slums for the destitute. T. H. Huxley pointed out in his essay "The Struggle for Existence in Human Society" that "The first men who substituted the state of mutual peace for that of mutual war . . . obviously put a limit upon the struggle for existence." But in statements which have been often quoted, he also acknowledged, "From the point of view of the moralist the animal world is on about the same level as a gladiator's show. . . . The strongest, the swiftest, and the cunningest live to fight another day . . . no quarter is given."

Many sociologists and naturalists revolted against the concepts of struggle for existence and laissez-faire philosophy, emphasizing that co-operative behavior among related living forms in nature was more common than active warfare. Darwin himself apparently did not share the simple-minded view that devouring or being devoured was the essential mechanism of selection and of evolution. In *The Descent of Man* he pointed out that "in numberless animal societies, the struggle between separate individuals for the means of existence disappears; struggle is replaced by co-operation." Samuel Butler and the Russian political philosopher Kropotkin went still further and promoted the view that co-operation and mutual help, rather than struggle, had been the principal agents of evolution. Kropotkin in particular became the standard-bearer of this attitude through his book, *Mutual Aid, A Factor of Evolution,* published around the turn of the century. During his travels in Siberia and Manchuria he had observed the frequent occurrence of co-operative endeavors in the animal world. The historical study of human institutions had convinced him furthermore that a desire for mutual aid had always prevailed among men Many observers since Kropotkin have provided examples of co-operation among animals, hardly compatible with the popular view of the law of the jungle.

Aggressiveness of course still continues to reign in the world of men, but it is probably true nevertheless that social mores have evolved along with the biological outlook. Whatever the actual practices in international and national politics, imperialism and social injustice are now universally denounced as contrary to moral and natural laws Tennyson would not today extol "Nature red in tooth and claw"; he would celebrate instead the co-operative ventures that men desperately try to develop in their home environment and on the international scene. In the part of Chicago known as "the jungle" Kropotkin's autographed portrait now presides over the activities of the Hull Settlement House founded by Jane Addams in the belief that "mutual aid" would transform this human jungle into a peaceful area. On a wall opposite the UN building New York advertises that swords will at last be converted into plowshares in the One World!

Idealist as he was, Kropotkin did not assume that the mutual aid of which he perceived so much evidence in nature was the result of a co-operative spirit based on sentiment. He traced it rather to an instinct of preservation which gave to species and to individuals a better chance of survival under conditions of stress. He realized that practically all living organisms eventually serve as food for others and that many individuals must perish in order that the group survive, but he also believed that aggressive warfare is unusual under normal conditions and that the wisdom of nature points to the advantages of peaceful coexistence. The present Zeitgeist in human affairs and in general biology has been further influenced by the increasing awareness that all living things are mutually interdependent. Everywhere in nature there is evidence that the different forms of life exist in an equilibrium which, although never stable, is nevertheless adequate under ordinary circumstances to permit the survival of most species. While altruism may have no part in the interrelationships among living things, it is certain at least that selfish interest is best served by opportunistic tolerance.

To a large extent, ecological equilibria are controlled by the availability of food and, as most creatures live at the expense of some other, the chain that binds all forms of life is, in final analysis, forged out of dead bodies Many curious facts have come to light illustrating the variety of the links which make up the chain of life. The records of the Hudson Bay Company show, for example, great fluctuations at intervals of years in the numbers of fur animals caught by trapping. These fluctuations reflect the availability of rodents on which the fur animals feed and the size of the rodent population is in turn controlled by the abundance of crops, by the weather, and by disease.

Likewise, the fish population fluctuates qualitatively and quantitatively in ocean waters under the influence of varied factors which affect the aquatic flora and fauna. The abundance of plankton and the kinds of plant and animal organisms prevailing in it change continuously from season to season and from year to year. When the current from the Atlantic predominates, it brings to the English coast water rich in phosphate, which favors one species of glassworm, and the result is a good herring year in Plymouth. In contrast, current from the Channel brings in water poor in phosphate, resulting in the predominance of other animal species and in failure of the herring fishery. Over Georges banks there occur now and then conditions which carry the young fish larvae into deep water to the south where they die—with the result that a poor haddock season can be predicted for New England waters. The "red tide" which swept immense numbers of dead fish onto the Florida beaches in 1946 and again in 1952, as it does approximately once a decade, was caused by a microscopic flagellate which secretes into the water a poison toxic to all marine life. The flagellate is always present in the waters off the Florida coast, but in numbers too small to be harmful; its population reaches toxic levels only when atmospheric circumstances bring about the local stagnation of low-salt brackish water in certain areas. Thus, all over the world, water movements affect the fish population indirectly but profoundly by controlling the abundance of small plant and animal life which acts either as source of food or as cause of disease for the larger species that feed on them.

Man and the World of Microbes

Man occupies a unique position in the pecking order of nature. Not only is he able to use anything that he wants as source of food, but under ordinary circumstances he need not fall prey to any living creature of another species. Prehistoric paintings and carvings show that he had even then achieved mastery over the most ferocious animals. There is only one significant exception to his biological dominance, but a very large one. Like any other living thing, man can become the victim of microorganisms, and in fact these account for a large percentage of his diseases. Furthermore, much of his economic life is under the control of microbial activities over which he has little if any control. Microorganisms add to the fertility of soil by converting plant debris into humus, but they also destroy a large percentage of the crops. They contribute essential steps to many technological processes, but they also spoil or rot every kind of foodstuff and of goods. Except in a few situations, micro-organisms are today as undisciplined a force of nature as they were centuries ago.

It is probably because man has so much less control over the microbial world than over the rest of life that micro-biological sciences often follow a course outside the main channels of modern scientific thought, and tend to be dom-inated by a mode of thinking which often appears naive in the light of modern biology. A few microbes are con-sidered *good* because they do things which give satisfaction to men, and most others have a *bad* reputation because their effects are considered detrimental in human affairs. This anthropocentric basis of judgment is of course philo-sophically questionable; moreover, its meaning is far less clear than appears at first sight.

The extent to which a microbe is good, bad, or indifferent is not always easy to decide even from the limited point of view of human interests, because these interests differ so widely from one type of circumstance to another. The same bacteria, yeasts, and fungi that are used by the Anatolian shepherd, the Bulgarian peasant, and the French farmer to convert milk into kumiss, yoghurt, or Camembert cheese are regarded by the sanitary American as objectionable be-cause they sour or putrefy his pasteurized dairy products. The viruses of German measles and of poliomyelitis, which used to be of trivial importance, are now becoming increas-ingly dangerous in the Western world as a result of im-proved sanitation. In these two cases, social habits which have not affected the viruses themselves have rendered them grave threats to human welfare.

Fashions, also, affect human attitudes toward certain kinds of infection. Ever since tulipomania became a national craze in Holland during the seventeenth century a few types of tulips have been highly valued because of the un-usual patterns of pigments in their blossoms. It is now known that these characteristics of the flowers are the re-sults of infection with a specific virus. For centuries Dutch and European bulb producers have carefully husbanded these types of tulips and thus have maintained the virus infection going in their stocks. Today fashions are changing, and the public has been educated to demand varieties of tulips which are more robust although less varied in flower pattern. The virus infection which was sought after in the past because it produced the delicate Zomerschon or Rembrandt tulips is now regarded as a disease which must be eliminated because it decreases the vigor and uniformity of the plants. Thus, it is clear that microbes are not good or bad in themselves, and that their relations to other living things are determined by circumstances independent of their own characteristics and propensities—including the whims in human taste.

Scientists investigate microorganisms in all their different facets just as they would any other form of life. The description of some cellular appendage or the analysis of a biochemical process is for some of them a sufficient end unto itself—with the kind of reward that a disinterested spectator or a curious child would enjoy. Another aspect of the study of microbial life is the ecological role that microorganisms play as a constant part of the environment—their complex interplay with the rest of the living world. Many scientists focus their interests on the relation that microbial activities bear to human welfare—an attitude which also is compatible with the highest criteria of theoretical achievement, as well demonstrated by Pasteur.

The knowledge that microorganisms can be helpful to man has never had much popular appeal, for men as a rule are more preoccupied with the dangers that threaten their life than interested in the biological forces on which they depend for a constructive existence. The history of warfare always proves more glamorous than accounts of co-operation. Plague, cholera, and yellow jack have found their way into the novel, the stage, and the screen, but no one has made a success story of the useful role played by microbes in the intestine or the stomach. Yet the production of much of the food that reaches our dinner table depends upon the activity of bacteria.

"The cow does not eat much," wrote the English boy in his school essay, "but what it eats it eats twice so that it gets enough." What the schoolboy really meant of course is that in the cow and other ruminants the rough components of grass are first acted upon by a whole variety of bacteria in the rumen and thus converted into soluble products which can be assimilated as food. Nutritional partnerships between the microbial world and animals take many forms. Bush sickness of sheep in Australia is one example of economic importance, which appears in animals pastured on soils lacking cobalt. Cobalt is not needed directly by the sheep, but lack of it interferes with the activities of certain bacteria which normally synthesize vitamin B_{12} in the gut. In other words, the sheep evolved to fit an ecological environment that must include the presence of cobalt and of certain types of bacteria which supply the vitamin B_{12} that the animal cannot manufacture itself.

Man is no exception to this rule of partnership with microbial life. The billions of bacteria normally present in the different sections of his intestine play a role as yet ill defined, but certainly important in manufacturing vitamins and other factors essential for his well-being. It is through the activity of the microbes in his gut that man has so long remained independent of chemical industry for his requirement in certain vitamins. Even lower animals depend on microorganisms for their survival. Termites owe their power of destruction to the microscopic protozoa in their gut which digest the cellulose of wood and thus allow them to derive nourishment from the timber in our dwellings.

The partnership between the microbial world and higher organisms has been investigated chiefly from the nutritional point of view, but it presents many other aspects at least as important. Thus, it is well known that the intestine in breast-fed infants contains almost exclusively a certain kind of bacteria (*Lactobacillus bifidus*), very different from the more common bacteria which predominate when the diet consists of cow's milk. This particular lactobacillus almost certainly plays a role in the greater resistance of breast-fed infants to enteritis during the early period of life. The female genital tract also contains normally certain types of bacteria which have beneficial effects.

Many other protective mechanisms involving the participation of the normal microbes of the body will certainly come to light as attention is focused on this problem. Oddly enough, existence of some of these protective mechanisms has become apparent from the fact that they almost disappear in patients treated with antimicrobial drugs. It is commonly observed that drug treatment brings about certain pathological disorders which indirectly result from elimination of the bacteria which normally play a useful role in the tissues. It is certain also that disappearance of one kind of microbial population is rapidly followed by the development of another kind which brings in its train unexpected and often dangerous consequences. The difficulties that may follow antibacterial therapy are in fact similar in essence to those encountered in any attempt to control predators in nature. Because so many leopards have been killed in Africa there is now a plague of baboons which destroy the crops in certain areas. When rabbits became scarce in England following introduction of the myxoma virus, the foxes and birds of prey began to raid poultry yards. On the Kaibab Plateau extermination of the wolves and mountain lions has proved unfavorable in the long run to the deer by allowing them to multiply excessively and overgraze their feeding areas. Whether the method of treatment affects the animal predators in the wilderness or the bacteria in the gut, it is always risky to tamper with the natural balance of forces in nature.

Infection versus Disease

Pasteur's first paper on the germ theory, the *Mémoire sur la fermentation appelée lactique*, appeared in 1857 just one month before Charles Darwin sent to Asa Gray the famous letter stating for the first time in a precise form the theory of evolution. Through this historical accident the germ theory of disease developed during the gory phase of Darwinism, when the interplay between living things was regarded as a struggle for survival, when one had to be friend or foe, with no quarter given. This attitude molded from the beginning the pattern of all the attempts at the control of microbial diseases. It led to a kind of aggressive warfare against the microbes, aimed at their elimination from the sick individual and from the community. No place here for the biological concepts now prevailing in other fields of natural history, according to which different living species can reach a modus vivendi compatible with their coexistence. The view that some sort of ecological equilibrium can be achieved between microbes and their potential victims has not been popular among physicians and medical scientists. In truth, it seems preposterous at first sight to speak of ecological equilibrium between the microbial agents of disease and the plants, animals, and men on whom they prey. What history records is not peaceful coexistence, but rather disastrous epidemics which surpass in destructive effects the most extreme attempts of man at total warfare.

In the preceding chapter we have seen that the chestnut blight provides an example of massive and sudden destruction of plants by microscopic fungi. Similar disasters are also common among animals. To mention only one, wholesale deaths of rats by plague were commonly illustrated during the Renaissance and heralded the appearance of the disease among men. Time and time again in history visitations of the plague bacillus have caused such immense mortality as to have constituted the most destructive cataclysms ever to strike mankind. Plague killed one half or two thirds of the inhabitants of certain cities of the Roman world—including England—during the Justinian era. It struck Europe again on several occasions between the fourteenth and seventeenth centuries with the same destructive effects. The outbreak of pneumonic plague in Manchuria half a century ago illustrated once more its tremendous killing power. Influenza, yellow fever, smallpox, typhus, and cholera likewise conjure up the thought of visitations descending on mankind as the curse of some avenging deity. Other epidemics are now forgotten, but caused great terror in their days. Such were the mysterious outbreaks of English sweating sickness, which ran a short but extremely acute

course during Tudor times, never to return again, at least
not in a recognizable form. Cataclysmic epidemics have
continued to occur here and there during the past century.
Most of them have been rather limited in area, exhibiting
great killing power only in populations newly exposed to a
particular species of microbe or placed under conditions of
great physiological misery. That epidemics can still strike
far and wide was proved during 1918 and 1919 when several consecutive waves of influenza killed some 20 million
persons—far more than had died as a direct result of the
preceding five years of unrestricted global warfare.

The spectacular decrease in the mortality caused by infections during the past century bears testimony to the effectiveness of the measures aimed at eradication of microbes. In reality, however, the role of these measures may
not have been so great as commonly believed. The toll of
human lives exacted by infection had begun to decrease
several decades before control measures inspired by the
germ theory were put into effect and almost a century before the introduction of antimicrobial drugs. As we shall
see, there came into play in Western Europe during the
second half of the nineteenth century biological and economic forces that increased the resistance of the social body
to infection. Granted the obvious usefulness of sanitary
practices, immunological procedures, and antimicrobial
drugs, it does not necessarily follow that destruction of microbes constitutes the only possible approach to the problem of infectious disease, nor necessarily the best. A century
ago it was thought on the western frontier that the only
good Indian was a dead Indian. Yet no one doubts today
that the white man and the Indian can coexist peacefully
and derive much mutual benefit from each other.

It is rarely recognized, but nevertheless true, that animals
or plants, as well as men, can live peacefully with their
most notorious microbial enemies. *Phytophthora infestans*,
the very fungus that caused the disastrous potato blight in
the 1840's, is still prevalent over the potato fields in Ireland
and in the rest of the world. But man has learned to use
farming methods that permit the potato to thrive even

though the fungus be present. Every year millions of white mice are raised in sanitary environments to meet the needs of laboratory science. Yet all these sleek-looking animals carry in their organs hosts of bacteria and viruses which are potentially capable of killing them but cause disturbances only when other things go wrong in the breeding colony.

Similarly, almost all human beings, Americans included, become infected with a host of microbial parasites of one sort or another. Bacteria such as tubercle bacilli, streptococci, or staphylococci, many types of viruses potentially capable of producing influenza, intestinal disorders, or various forms of paralysis, all kinds of protozoa and worms, are commonly present in the tissues of individuals who consider themselves hale and hearty. A generation ago each and every person lived in almost constant daily contact with tubercle bacilli, became a little tuberculous, yet had a fair chance of enjoying a normal, creative life. In brief, the presence of pathogens in the body can bring about disease, but usually does not. The world is obsessed—naturally so—by the fact that poliomyelitis can kill and maim several thousand unfortunate victims every year. But more extraordinary, even though less dramatic, is the fact that millions upon millions of young people become infected with polio viruses all over the world, yet suffer no harm from the infection.

Thus, while many types of microbes can paralyze, starve, or bleed their victims, and are endowed with the power to kill them within a few days or a few years, it is also true that the same microbes are usually harbored for a whole lifetime by normal, very ordinary citizens, who are not even aware of being infected and who—for all we know—may derive some unrecognized benefit from their infection. The dramatic episodes of conflict between men and microbes are what strikes the mind. What is less readily apprehended is the more common fact that infection can occur without producing disease.

Adaptive Mechanisms of Resistance to Infection

Men, animals, and plants normally possess mechanisms
that permit them to resist infection, and it can be assumed
that this resistance, which is essential to survival, is acquired through the processes of evolutionary adaptation. Its
mechanisms are multiple and differ from case to case, but
in general outline they are not unlike the processes involved
in adaptation to the physicochemical environment (Chapter II). As we have seen, some of the adaptive mechanisms
correspond to hereditary changes resulting from the selection of mutant forms of the host best fitted for survival.
During a widespread epidemic of tuberculosis, for example,
the most susceptible are likely to die young, leaving no
progeny. In contrast, many of those who survive are geneti-
cally endowed with a high level of natural resistance which
they pass on to their descendants. The low tuberculosis
mortality prevailing in the Western world at the present
time is in part the result of the selective process brought
about by the great epidemic of the nineteenth century
which weeded out the susceptible stock. It is worth pointing out in passing that this selective mechanism can be
effective only against diseases affecting the individual before and during the reproductive age. The fact that selection is less likely to operate for diseases occurring later in
life may greatly add to the difficulty of developing methods
of control for the afflictions characteristic of old age.

In addition to genetic mechanisms of resistance there are
nonhereditary protective reactions that occur in the individual as a response to contact with infectious agents. Under the conditions that used to be regarded as natural not
so long ago, before the fear of disease became an obsession
and sanitary living a religion, all children were exposed
early in life to many of the pathogens common in their environment. While many died of the infections thus contracted, those who recovered acquired thereby an immunity which served them well throughout the rest of their
lives. As recently as a generation ago it was accepted as a

law of nature that children had to go through the fire of certain diseases then regarded as peculiar to childhood. The reason that adults were usually resistant to these so-called childhood diseases was simply that they had once overcome their onslaught during youth. This was a cruel way of achieving adaptation to microbial agents of disease, but it was nature's way, expensive yet effective. In a not too-distant past children were expendable. Many more were produced than were needed to keep the species going, a situation which still prevails in most of the world. As a result of effective birth control and of changed ethics, however, it is now desired that all children survive. This ideal naturally makes it essential to discover ways that will substitute for the protective effects that childhood diseases used to exert in the past. Vaccination to elicit without danger the immunity formerly resulting from the disease itself and sanitary procedures designed to eliminate infectious agents from the environment are the technological procedures through which man attempts to substitute for the natural mechanisms of biological adaptation.

The adaptive processes which increase resistance to infection are an important aspect of ecological equilibrium between microbes and their potential victims. For example, while plague is highly fatal for rats in the Western world, the plague bacillus is almost ubiquitous among the rats of Bombay, causing in them an infection so mild as to be often inapparent. Countless generations of contact with the bacillus have clearly called into play in the Bombay rat adaptive mechanisms which limit the harmful effects of infection. The ebbs and flows of human diseases in the course of history suggest that they, too, are under the control of natural selective mechanisms.

An interesting testimony to this fact is given by Fracastoro in his famous poem "Syphilus" published in 1548, which gave its name to syphilis. The disease had burst upon the European stage in February 1495, when the French army entered Naples. From the contemporary descriptions it was then an acute and terrifying experience very unlike the much milder forms seen in our times. Fracastoro states

that he wished to describe syphilis in precise details in order to permit its recognition when it would return later
with its original severity. For, according to him, it had progressively lost in ferocity since first striking Italy barely half
a century before publication of the poem.

During the eighteenth and nineteenth centuries the
Amerindian and Polynesian populations were decimated by
epidemics of smallpox, measles, tuberculosis, and other infections contracted from European explorers and invaders.
After a few generations, however, the mortality caused by
these diseases began to fall spontaneously and progressively.
Some of the historical facts concerning the evolution of
these epidemics are sufficiently well documented to warrant description in some detail.

Smallpox was apparently introduced in the American
continent early during the Spanish Conquest—probably by
a Negro in Cortez' band. The Indians proved highly susceptible to the disease, which almost wiped out some of
their settlements—and there is reason to believe that this
disaster contributed to their rapid defeat by the Spaniards.
The conquest of North America a century later provided
further dramatic evidence of the susceptibility of the Amerinds to smallpox. Repeated outbreaks decimated village
after village, and at times whole tribes. Early in the seventeenth century, for example, the Massachusetts and Narragansett Indians were reduced in a short time from 30,000
and 9,000, respectively, to a few hundreds. In the epidemic
of 1837 the Mandan population fell from 1,600 to 31, the
Assiniboins lost whole villages, the Crows one third of their
population, while the total deaths among Plains tribes
amounted to 10,000 in a few weeks. Similar outbreaks occurred in 1870–71 among the Blackfeet.

The inhabitants of Tierra del Fuego suffered even more
from the diseases brought in by the Europeans. The Ya-
mana Indians, hardy people engaged in fishing, were reduced from approximately 3,000 at the time of Darwin's
visit to less than 1,000 in 1884; when measles swept the
territory the tribe was reduced to 400. Many more deaths
occurred during the following years, leaving only 170 sur-

vivors in 1908. The other tribes suffered a similar fate, with the result that from 9,000 Indians in 1848 the population of Tierra del Fuego had been reduced to 150 purebreds in 1947.

Microbial diseases among the Polynesians during the past century present a picture similar to that seen among the Amerinds. The South Pacific and Hawaii were explored during the second half of the eighteenth century. Granted that the charm of the Pacific Islands and the amorous welcome of their women may have warped the judgment of the European visitors, there is no reason to doubt the validity of their unanimous opinion concerning the health and vigor of the Polynesian people at that time. European navigators saw in all the islands robust and happy men and women obviously well adapted to their environment—including their local pathogens. Yet disease became rampant among the islanders within a short time after these early explorations and the Polynesian population soon began to fall From approximately 300,000 at the time of Cook's first visit in 1778, the population of native Hawaiians had fallen to less than 37,000 in 1860. During the same period the population of the New Hebrides was reduced to one tenth of its original size.

There is no doubt that this holocaust was brought about largely through the venereal diseases, tuberculosis, scarlet fever, measles, and other infectious disorders that the Polynesians contracted during their short contacts with the Europeans. When measles was introduced into Hawaii practically the whole population was stricken with the disease and many thousands died. Measles also took the lives of King Kamehameha II and Queen Kamamalu during their visit in England, the king and queen being stricken shortly after their arrival in London in July 1824. Sir Henry Halford, president of the Royal College of Physicians, and the two other attending English doctors found it difficult to believe that a disease "which even a delicate London girl might bear could be so destructive to robust denizens of the Pacific." Epidemics of measles, pertussis, and influenza struck Hawaii in 1848 and every child born that year died.

In 1853 there were over 9,000 cases of smallpox with 6,000 deaths out of a population of 70,000. Cook himself was aware of this fact. During his second and third visits to the South Pacific he was much disturbed at the thought that his sailors were responsible for having introduced venereal diseases into Tahiti; he found solace, however, in the belief that the initial guilt was to be placed on Bougainville's French crew.

The destructive power of tuberculosis when it first strikes a population is well illustrated by the epidemics among the Plains Indians of western Canada. In the 1890's the annual death rate from tuberculosis in the Qu'Appelle Valley Reservation of Saskatchewan reached the fantastic figure of close to ten per cent of the total population. More than half the Indian families were eliminated in the first three generations of the epidemic. Moreover, some twenty per cent of the deaths in the surviving families were caused by tuberculosis.

Many other epidemic outbreaks of disease could be selected to illustrate the destructiveness that viruses and bacteria often exhibit when first introduced into a population. Several enlightening examples have occurred recently, such as the outbreak of measles among the Tupari Indians of the Brazilian forest. This small tribe had remained isolated and apparently without contact with white people until 1949, when first discovered. At that time it counted some 200 people. When revisited six years later, in 1955, two thirds of the group had died of measles, caught from rubber gatherers. Measles also struck the Eskimos of the Canadian arctic in 1952; the attack rate in this instance was over ninety-nine per cent, including all ages, and the mortality rate reached up to seven per cent at Ungava Bay. An outbreak of poliomyelitis among a group of Eskimos of the Hudson Bay in 1949 exhibited the same pattern. Fourteen per cent of the population died and over forty per cent developed paralytic disease, all age groups being involved. Still another example has been made familiar by the writings of Albert Schweitzer. Sleeping sickness (African trypanosomiasis) is a new disease in the Ogowe region of Equatorial Africa. It was introduced some thirty years ago by carriers that came with the Europeans from Loango, where it has apparently existed from time immemorial. The disease proved terribly destructive in its new territory, carrying off one third of the population in the course of a few years. In Uganda it killed 200,000 persons out of 300,000 in six years. Of the 2,000 inhabitants in a village of the Upper Ogowe, only 500 survived after two years of the epidemic.

The examples cited—and these are but a few among many—leave no doubt that whole populations can be decimated by pathogens with which they have had little contact in the past. But it is becoming apparent, on the other hand, that the diseases introduced by the white man in the eighteenth and nineteenth centuries no longer exhibit among primitive peoples the very acute course with a rapidly fatal outcome which was uniformly observed in the past. These people have developed, or are developing, a manner of biological response to infection which is similar to that seen in the Western world under normal conditions. The high mortality rates caused by the plague bacillus or by the yellow fever virus among Europeans can probably serve as an analogy to the type of virulence that the tubercle bacillus or the measles virus had for the Polynesians or the Amerinds two centuries ago. Tuberculosis and measles are still important causes of disease among these people but rarely give rise to acutely devastating epidemics in the areas where they have been established for several generations. Similarly it has been observed that African trypanosomiasis loses after a while the terrific virulence that it exhibits when first introduced into new districts of Equatorial Africa. The disease lingers on, but it carries off small numbers of victims instead of killing two thirds of the exposed population as it once did.

Precise observations are available concerning the evolutionary changes in the clinical manifestations of tuberculosis among the Indians of North America. During the first and second generations of the tuberculosis epidemic on the Qu'Appelle Valley Reservation, acute forms of the disease were very frequent—evidence of extreme susceptibility of the host. In 1921, at a time when the generalized epidemic was in the third generation, the disease showed a greater tendency to localize in the lung and to exhibit a chronic course, the mortality was falling, glandular involvement had dropped to seven per cent among Indian school children. This latter manifestation of high susceptibility to the disease has continued to decline steadily and is now seen in less than one per cent of children in the present (fourth) generation. In other words, while tuberculosis among the Amerinds exhibited at first a very acute course, different in character from that observed in people who have had contact with the tubercle bacillus for several generations, it is now undergoing a change which makes it resemble the more chronic type of disease commonly seen in the Western world under conditions of social stability. There is no doubt that a similar evolutionary process took place among many groups of Caucasian people many generations ago. Tubercle bacilli became ubiquitous in European and American cities during the nineteenth century, and the continuous elimination of the human stocks most susceptible to them left a population endowed with a significant degree of natural resistance to tuberculosis.

Striking as it is, the epidemiological evidence that evolutionary adaptive processes play a decisive role in resistance to disease fails to carry conviction—especially for those reluctant even to consider the possibility that natural forces are more effective than medical procedures in the long run. Fortunately, the evolutionary theory of resistance may soon be demonstrated by the results of an extraordinary experiment on a continental scale which has been going on during the past decade.

Rabbits did not exist in Australia until 1859; that year Thomas Austin introduced twenty-four animals from Europe, just in the interest of sport, and six years later he counted more than 30,000 rabbits on his estate. In fact, the land and climate of Australia proved so favorable to the rabbits that they multiplied at an extraordinary rate, eventually reaching numbers in the billions and destroying crops and pastures. Hunting, trapping, and poisoning proved without avail against the new plague. So great was the destruction of crops that the government of New South Wales in August 1887, offered a prize of £25,000 to anyone demonstrating an effective method of extermination. Immediately after reading this announcement in the newspapers Pasteur suggested, in 1887, that one might eradicate the rabbits by contaminating their food supply with the proper kind of bacteria. In fact, he went so far as to run a preliminary field test on the Pommery Estate in the Champagne country, spreading virulent cultures of *Pasteurella multocida* on alfalfa around burrows. Although the test resulted in the death of many rabbits, practical reasons made it impossible to do anything further with the technique. It is now recognized, indeed, that the type of bacteria selected by Pasteur could not have given rise to a progressive epidemic.

In a modified form Pasteur's suggestion was successfully put into effect a few years ago. In 1950 a highly virulent strain of myxomatosis virus was introduced into Australia and it rapidly became established in the rabbit population through the agency of mosquitoes, which carried it across a large part of the continent. Two years later, release of the virus on a private estate in France (where it did not exist either) similarly resulted in its spread over much of Continental Europe and England. Myxoma occurs naturally in the wild rabbits of Brazil, but merely in the form of a benign tumor, which does not endanger the life of the animal The mild character of the infection in the Brazilian rabbit clearly denotes an ancient association with the myxoma virus, resulting in ecological equilibrium. In contrast, the virus recovered directly from these animals causes an acute, almost uniformly fatal disease when inoculated into the rabbits of Europe or Australia, which have never been exposed to the disease

The initial outbreaks of myxomatosis in Australia were characterized by an enormously high case mortality rate—higher than ninety-nine per cent. Within a year, however, the case mortality had fallen to ninety per cent in areas where a second spontaneous outbreak had occurred. This fall in mortality was due in part to a decrease in the virulence of the virus. In Australia the virus is transmitted from rabbit to rabbit almost entirely through mosquitoes which act mechanically as "flying needles." Because the highly virulent strain of virus killed the rabbits within a very few days, the chances for its transmission through the mosquito vector were rather limited. However, when a virus of lower virulence appeared spontaneously by mutation, it produced in the rabbit a less rapidly fatal disease, with skin lesions of longer duration. Thus, the less virulent mutant strain had a better chance of being transmitted through mosquito bite and it progressively displaced in the field the original highly virulent strain.

The evolution of rabbit myxomatosis in Australia has also provided evidence that hereditary changes in the host are likewise of importance in altering the course of epidemics. As already mentioned, the European type of rabbit introduced into Australia is immensely susceptible to myxomatosis. As a result, the mortality among infected animals proved to be almost total the first year that the virus was successfully released in Australia and Europe. Recently, however, it has been found that rabbits trapped in areas of these countries where the infection has taken hold exhibit a much higher degree of resistance to the most virulent forms of the virus than that exhibited by rabbits before the beginning of the epizootic. It has been established, furthermore, that this increase in resistance has a genetic basis and must therefore result from the selection of mutant animals which had survived the initial infection.

The touchstone of a scientific theory is its power to predict natural phenomena, but evolutionary theories can rarely be put to this test because most biological processes evolve so slowly. There are indications, however, that rabbit myxomatosis will evolve rapidly enough in Australia and in Europe to provide convincing evidence that epidemics spontaneously decrease in severity through adaptive changes affecting both the host and the parasite. It seems justified to predict that the rabbit and the virus of myxomatosis will eventually achieve in Australia and in Europe an ecological equilibrium that will result in the survival of both. Just as the Bombay rat has become resistant to plague, the rabbit will become relatively resistant to myxomatosis. And if the rats and rabbits have a racial memory they may come to take pride in the illusion that it was through some conscious action of their own that they achieved control over the great epidemics of their past, just as men believe that medical practices have been the only factors of significance in the control of leprosy, measles, scarlet fever, syphilis, tuberculosis, or any number of infections caused by microorganisms ubiquitous in human communities.

Symbiosis and Parasitism

As we have seen, it is a universal biological law that living things constantly harbor a host of different kinds of microbes, some of which contribute to their well-being while others become the cause of disease under certain circumstances. The words "symbiosis" and "parasitism" have long proved useful to denote these contrasting relationships. But words are treacherous, allowing themselves to be applied in circumstances and to purposes for which they were not designed. Symbiosis and parasitism refer to certain types of relationships existing between two living things at a given time, but the words should not imply that one of the organisms involved in the partnership is of necessity and permanently a symbiont or a parasite—always behaving either as a useful or as a dangerous member of the partnership. Among men the need for defense against a common enemy often makes for strange bedfellows, and contrariwise the stresses and shortage of food in a concentration camp engender bitter competition among individuals devoted to each other's welfare in ordinary life. Similarly, at all levels of organization, the outcome of the interplay between two individuals is determined not only by their intrinsic endow-

ments but also, and even more, by the conditions under
which they come into contact.

Just as microorganisms known to be capable of causing
fatal disease can persist in the body without manifesting
their presence, so can others regarded as useful become destructive when circumstances change. Symbiosis is usually
on the threshold of disease. The profound effects that the
environment can exert on the interrelationship between
microorganisms and their hosts are well illustrated by the
behavior of the nodule bacteria in legume plants. On the
rootlets of beans, peas, and other legumes there exist swell-
ings of various sizes which are known as nodules, and which
are the tissue response of the plant to the presence of certain special kinds of bacteria living in association with its
root system. The nodule bacteria derive much of their
nourishment from the plant, and in exchange they make
available to the plant soluble nitrogenous compounds that
they snythesize from the nitrogen of the air—a perfect example of symbiosis mutually beneficial to both partners.

Several physiological adjustments assure the success of
this type of symbiosis under usual circumstances. Thus, establishment of the nodule bacteria is facilitated by the production in the root of a kind of exudation at a certain stage
in the germination of the plant. Modifications of both
bacteria and host tissues occur soon after infection, the root
nodules becoming a modified root adjusted to the requirements of the bacterial symbiont. Since excessive production
of nodules would deprive the plant of its functional root system and then terminate its existence, the association is regulated by inhibitors produced in the meristems of the roots
and nodules. However, these highly organized and mutually beneficial relationships can be readily upset. When the
plant is grown in a soil or a medium deficient in boron, the
bacteria no longer fix nitrogen and, furthermore, they attack the protoplasm of the plant. From symbiotic their behavior thus tends to become parasitic, even though their
intrinsic characters are not changed by the absence of
boron. As mentioned earlier, similar changes from symbiotic relationship to parasitic behavior occur in other

biological systems including man. Suffice it to mention that the bacteria of the intestine, which serve useful functions under normal circumstances, can under others become responsible for toxic reactions and even cause death.

The concrete facts of microbiological sciences have been on the whole easy to discover and their understanding presents no abstruse problem. The real difficulty has been rather to explain why so many varieties of microorganisms, endowed with the ability to kill, usually produce only self-limiting disease processes and often cause no discernible harm even though they persist in the body. True to Goethe's remark that where understanding fails there immediately comes a word to take its place—many words have been used to deal with this apparent anomaly. Abstract concepts such as resistance or susceptibility of the host, virulence, attenuation, invasiveness, and toxicity of the parasite have all been invoked as if they corresponded to real entities, and could be incorporated into a formula to account for the outcome of infectious processes. Unfortunately, these words refer not to characteristics inherent in the host or the parasite but merely to states of relationship between the two And these relationships are governed by biological properties that still transcend description in the conventional terms of present-day physicochemical sciences.

For the sake of scientists who tend, or pretend, to despise abstract concepts and to have respect only for hard facts, it is worth pointing out that the most abstract formulation of the problem of parasitism first came from the American pathologist Theobald Smith, who more than any other American scientist contributed important hard facts to microbiological knowledge. In his classical essay on Parasitism and Disease, Smith suggested that it is of biological advantage for the parasites not to kill their hosts, since disappearance of the host jeopardizes the parasite's survival. The most successful parasite, in other words, is the one that allows to its victim as much life as is compatible with its own needs. In reality, as we have seen, the equilibrium between microorganism and host that is implied in successful parasitism is rarely stable. Microorganisms capable of persisting for prolonged periods of time in the body in the form of inapparent infections may all of a sudden undergo unrestricted multiplication, with disastrous effects for their host. Thus, the concept of successful parasitism corresponds to a statistical statement valid only for a population as a whole, but in each particular case the outcome of the relation between a given microorganism and a given host is determined by the special circumstances under which the two come into contact.

Ecological equilibrium with microorganisms is an ideal state but one which is not readily achieved and is frequently disturbed. Microbial diseases are the manifestations of its failures. Some of the cataclysmic outbreaks of history can be readily traced to situations in which evolutionary mechanisms have not had a chance to operate. For example, the introduction of an infectious agent new to a community often gives rise to widespread epidemics, affecting simultaneously large numbers of individuals. The chestnut blight on the American continent, rabbit myxomatosis in Australia and Europe, smallpox, measles, tuberculosis, when they first struck the Indians, the Polynesians, or the Eskimos, have during the past century provided spectacular examples of the tremendous susceptibility of populations to infectious agents new to them, before ecological equilibrium had been reached. It is probable also—but as yet unproved —that infectious agents can undergo mutations endowing them with a new property to which the populations in which they exist are not adapted. Such mutants can in theory acquire the destructive potentialities of new invaders. The influenza pandemic of 1918–19 may have been caused by a strain of the influenza virus slightly different from the forms then widely distributed among hogs and men.

In our communities the situation is very different for infections such as measles, rheumatic fever, tuberculosis, osteomyelitis, the various respiratory illnesses, and for so many other diseases caused by viruses and bacteria which have been ubiquitous for countless generations. In these cases infection is extremely widespread but disease is only an accidental occurrence, affecting a very small percentage of infected individuals. In general the adaptive relationship between microorganism and host is effective only for the precise circumstances under which adaptation evolved—circumstances which constitute physiological normalcy for the host concerned. Any departure from this normal state is liable to upset the equilibrium and to bring about a state of disease. Unrestricted multiplication of the microorganism is then merely a consequence of the failure—even though transient—of the adaptive mechanisms of resistance.

Thus, there is beginning to emerge a general pattern which permits some generalization concerning the interrelationship between microbial pathogens and other living things. When a population—of plants, animals, or men—is exposed to a pathogen with which it has had no past experience, exposure may bring about severe disease in many of its individuals. The generalized epidemic, however, soon calls into play adaptive changes in both the host population and the infective agent which bring about an ecological equilibrium between them. The infective agent may remain widely distributed in the community, but its presence need not be associated with injurious effects. Disease, when it occurs, is due to a change in the conditions under which the ecological equilibrium had evolved. These changes may be of varied nature. In man the provocative cause of microbial disease may be a disturbance in any of the factors of his external or internal environment—be it weather conditions, availability of food, working habits, economic status, or emotional stress.

Pasteur had clearly visualized these complexities and had pointed out explicitly in his writings that the response of the infected individual was determined by his hereditary endowment, his state of nutrition, his environment including the climate, and even his mental state. In the course of studies on the disease of silkworms known as flacherie Pasteur came to the conclusion, startling for the time, that the microorganisms present in such large numbers in the intestinal tract of the sick worms were "more an effect than a cause of the disease." These words were echoed half a century later by G. B. Shaw's facetious remark in the preface to *The Doctor's Dilemma* "The characteristic microbe of a disease might be a symptom instead of a cause."

IV.

ENVIRONMENT AND DISEASE

The Weather, the Potato Blight, and the
Destiny of the Irish

In the epilogue to *War and Peace* Tolstoy attempted to justify the structure of his novel by contrasting the techniques used by the historian and by the artist in reporting political and social events. The historian works under the illusion, Tolstoy claimed, that he can deal with his material in a scientific manner and provide rational accounts of past situations. He pretends that his knowledge of background, circumstances, and participants permits him to explain the outcome of historical happenings. In reality, however, he selects and emphasizes only those determinant factors which conform to his prejudiced views of history. The artist has no such illusion. He does not pretend to act like a scientist and yet he presents of historical events a picture that is truer to reality. Instead of explaining history he evokes its complexity and subtleties by describing the atmosphere in which events took place and the emotional reactions of each individual participant.

In *War and Peace* Tolstoy labored the thesis that military commanders—Napoleon included—are passive instruments who register and exploit situations but do not determine their course. Wars, like all human affairs, are so complex in their determinism that they can hardly be accounted for by the ordinary processes of reason. Social forces, economic factors, personal ambitions, or political doctrines are not the real causative agents of history, and the feats or words of military heroes, statesmen, or philosophers influence its course even less. Men usually find themselves in circumstances which they cannot comprehend and over which they have no control. Since the real causes of phenomena, wrote Tolstoy, are hidden beyond the reach of the human mind, historians can at most describe the behavior of individuals and certain limited interrelationships, but they must abandon the futile search for the specific causality of human events.

Tolstoy wrote *War and Peace* between 1863 and 1869 and his skepticism with regard to historical causality was a reaction against scientific materialism. Ironically enough, however, his novel became immensely popular precisely at the time when the doctrine of specific causality was achieving its most spectacular successes and gaining almost universal acceptance in medicine. Nevertheless, Tolstoy had been somewhat of a scientific prophet. The difficulties encountered in determining the factors that made Napoleon invade Russia in 1812 have their counterpart in the failure to account for most of the phenomena bearing on health and happiness in terms of simple and direct cause-effect relationships. In fact, Tolstoy's view that historical events cannot be attributed to single causes applies to most situations in the world of nature. The story of the roundabout way in which a microscopic fungus probably native to Central America destroyed the potato crop in Ireland and exerted thereby a dramatic influence on the destiny of the Irish people, illustrates the complexity of the interplay between the external environment and the affairs of man.

As far as is known, the potato originated in the Andes, where it still grows wild, yielding tubers so small as to be hardly fit for human use. In its native habitat the plant is infected with the parasitic fungus *Phytophthora infestans* but suffers little, if at all, from its presence. Through evolutionary adaptation, the fungus and the wild potato have obviously achieved a state of ecological equilibrium which permits the survival of both. Eventually the potato was improved for human consumption, and it became one of the most important sources of food in the Western world after the eighteenth century. While the fungus phytophthora has followed the potato wherever the plant has been taken, the relationship between the two has changed, the improved varieties of potatoes selected for large yields being much more susceptible to infection than are the wild varieties. Fortunately, it is possible by proper techniques of farming to arrange that most of the potato crop escapes destruction by the parasite. Now and then, however, the weather conditions upset the best farming practices, the fungus multiplies more rapidly and abundantly than usual, and kills the plant.

The potato blight broke out on a disastrous scale in Europe and particularly in Ireland around 1845. For two years in succession the blight not only killed the foliage but rotted the tubers in the ground and in storage. Because the impact of the disaster was so varied and so great, it is worth recording in some detail the constellation of circumstances under which it occurred and the scientific debates to which it gave rise.

Weather had been very unpleasant shortly before the blight broke out. For several weeks the atmosphere had been one of continued gloom, with a succession of chilling rains and fog, the sun scarcely ever visible, the temperature several degrees below the average for the previous nineteen years. The botanist John Lindley held the theory that bad weather had caused the potato plants to become saturated with water. They had grown rapidly during the good weather, then had absorbed moisture with avidity when the fog and the rain came. As absence of sunshine had checked transpiration, wrote Dr. Lindley, the plants had been unable to get rid of their excess of water and in consequence had contracted a kind of dropsy. Putrefaction was the result of this physiological disease. The Rev. Miles Berkeley, a naturalist with much knowledge of the habits of fungi, held a different theory and connected the potato disease with the prevalence of a species of mold on the affected tissues. To this Lindley replied that Berkeley was attaching too much importance to a little growth of mold on the diseased potato plants. He added that "as soon as living matter lost its force, as soon as diminishing vitality took the place of the customary vigour, all sorts of parasites would acquire power and contend for its destruction. It was so with all plants, and all animals, even man himself. First came feebleness, next incipient decay, then sprang up myriads of creatures whose life could only be maintained by the decomposing bodies of their neighbours. Cold and wet, acting upon the potato when it was enervated by excessive and sudden growth, would cause a rapid diminution of vitality, portions would die and decay, and so prepare the field in which mouldiness could establish itself."

Thus, the professional plant pathologists, represented by Lindley, believed that the fungus could become established on the potato plant only after the latter had been debilitated by unhealthy conditions, whereas Berkeley saw the fungus as the primary cause of the disease, with fog and rain as circumstances which favored its spread and growth. In this manner the controversies which were to bring Pasteur in conflict with the official world of the French Academy of Medicine in the 1880's were rehearsed three decades earlier in the pages of the English *Gardener's Chronicle.*

It must be emphasized that the destruction of the crop in 1845 was not the result of a new infection by *Phytophthora infestans.* The fungus had been present on the potato plant since its introduction from Central America, but it took unusual weather conditions to render the plant highly susceptible to infection. Although the fungus persisted in Ireland after the Great Blight, it was only during occasional years that the weather was propitious for its proliferation, so that potato culture recovered progressively.

The two years of the blight, however, had been sufficient to ruin the economy of Ireland. Following the introduction of the potato during the eighteenth century, the Irish population had much increased, as is always the case when a new source of food becomes available. From 3.5 million around 1700 it had reached approximately 8 million in 1840. The potato blight caused an acute food shortage, with the result that a million persons died of outright starvation. Furthermore, many of those who escaped death became more susceptible to a variety of infectious diseases. Thus began a great epidemic of tuberculosis which after a century is only now beginning to abate. Also, lack of food and economic misery forced a large percentage of the Irish population to emigrate, particularly to America. Even today the population of Ireland is only half what it was before the potato famine.

In America the Irish immigrants found work in the mushrooming industries of the Atlantic seaboard. But they found also crowded and unhealthy living conditions. Coming from rural environments they suddenly experienced the worst aspects of slum existence. The profound upheaval in their way of life made them ready victims to all sorts of infection. The sudden and dramatic increase of tuberculosis mortality in the Philadelphia, New York, and Boston areas around 1850 can be traced in large part to the Irish immigrants who settled in these cities at that time.

Thus, all sorts of accidents played their part in linking tuberculosis—the Great White Plague of the nineteenth century—to a fungus living on the wild potato in Central America. The change in ecological relationship between fungus and potato that occurred when the latter was removed from its native habitat and was "improved" for human consumption; the disturbance in the internal physiology of the potato caused at a critical time by unusual weather conditions; the biological and cultural urges which brought about the rapid increase in the Irish population during the first part of the nineteenth century—all these forces and many social factors that cannot be discussed here played an essential part in transforming Pat the Irish pigtender into a New York City cop. If ever a writer succeeds in making a popular story of the potato blight, he may conclude, as Tolstoy did for Napoleon's invasion of Russia, that its determinism is beyond human analytical power. In fact, it is perhaps just an illusion of science to believe that the vagaries of the relations between the potato and a microscopic fungus, inadequate farming practices, and the weather conditions in the 1840's were the real factors that led the adventurous spirit of man to establish on the American continent the wit of the Irish, their Catholic faith, and their political genius.

The Doctrine of Specific Etiology

Until late in the nineteenth century disease had been regarded as resulting from a lack of harmony between the sick person and his environment; as an upset of the proper balance between the yin and the yang, according to the Chinese, or among the four humors, according to Hippocrates. Louis Pasteur, Robert Koch, and their followers took a far simpler and more direct view of the problem. They showed by laboratory experiments that disease could be produced at will by the mere artifice of introducing a single specific factor—a virulent microorganism—into a healthy animal.

From the field of infection the doctrine of specific etiology spread rapidly to other areas of medicine; a large variety of well-defined disease states could be produced experimentally by creating in the body specific biochemical or physiological lesions. Microbial agents, disturbances in essential metabolic processes, deficiencies in growth factors or in hormones, and physiological stresses are now regarded as specific causes of disease. The ancient concept of disharmony between the sick person and his environment seems very primitive and obscure indeed when compared with the precise terminology and explanations of modern medical science.

Unquestionably the doctrine of specific etiology has been the most constructive force in medical research for almost a century and the theoretical and practical achievements to which it has led constitute the bulk of modern medicine. Yet few are the cases in which it has provided a complete account of the causation of disease. Despite frantic efforts, the causes of cancer, of arteriosclerosis, of mental disorders, and of the other great medical problems of our times remain undiscovered. It is generally assumed that these failures are due to technical difficulties and that the cause of all diseases can and will be found in due time by bringing the big guns of science to bear on the problems. In reality, however, search for *the* cause may be a hopeless pursuit because most disease states are the indirect outcome of a constellation of circumstances rather than the direct result of single determinant factors.

It is true that in a few cases—far less common than usually believed—the search for *the* cause has led to effective measures of control. But it does not follow that these measures provide information as to the nature of the trouble that they correct. While drenching with water may help in putting out a blaze, few are the cases in which fire has its origin in a lack of water. The story of insulin and diabetes well illustrates that the discovery of a therapeutic agent does not necessarily solve the problem of disease causation.

Diabetes was first produced in experimental animals by interfering with pancreatic secretion, and this discovery led to the preparation from pancreas of a substance, insulin, which plays an important role in the metabolism of sugar. Insulin was then shown to be highly effective in the treatment of diabetes in man. This therapeutic triumph is probably the most elegant and spectacular achievement of medical science, but its bearing on the etiology of diabetes is far from clear. While diabetes can be produced in experimental animals by injuring the pancreas and thus interfering with the production of insulin, the disease as it occurs in man is a general metabolic dysfunction affecting the metabolism of protein, fat, and mineral, as well as of sugar. The primary disturbance may be in some part of the body quite remote from the pancreas and the deficiency of insulin may be secondary to it. Treatment with insulin corrects the manifestations of diabetes but it has no effect on the primary lesion of the disease, which remains unknown in many cases. Likewise, cortisone is highly effective against many inflammatory states which do not originate from a lack of this hormone in the patient, just as aspirin, which is a synthetic drug foreign to the body, can alleviate pains and headaches.

Thus, effective therapies do not constitute evidence for

the doctrine of specific etiology, and there are many cases in which a given disease can be controlled by several unrelated procedures. The incidence of malaria in a community can be reduced by drugs that attack the parasite, by procedures that prevent mosquitoes from biting man, by insecticides that poison the mosquitoes, or by agricultural practices that interfere with their breeding.

The difficulties inherent in the concept of causation of disease are now apparent even with regard to tuberculosis, long thought to have provided the most spectacular demonstration of the doctrine of specific etiology. All textbooks dealing with infectious diseases consider the discovery of the tubercle bacillus as the highest peak of the science of medical microbiology. The circumstances were indeed dramatic. At that time tuberculosis was by far the most important disease in the Western world. The tubercle bacillus was difficult to visualize by microscopic techniques, and even more difficult to cultivate *in vitro*. Yet Robert Koch succeeded in demonstrating its presence in all tuberculous tissue that he studied and in producing at will experimental tuberculosis by injecting small amounts of cultures of the bacillus into guinea pigs, rabbits, and mice. How could one doubt, after these spectacular achievements, that the bacillus isolated by Koch was *the* cause of tuberculosis?

There was, however, another aspect of the problem that had remained hidden from Koch. It can be stated with great assurance that most of the persons present in the very room where he read his epoch-making paper in 1882 had been at some time infected with tubercle bacilli and probably still carried virulent infection in their bodies. At that time, in Europe, practically all city dwellers were infected, even though only a relatively small percentage of them developed tuberculosis or suffered in any way from their infection. Koch himself was infected. When he injected tuberculin into his own arm in 1890 he suffered one of the most violent allergic reactions on record, evidence of the fact that the tubercle bacillus had at some earlier time multiplied in his body. But Koch did not have clinical tuberculosis, and he remained a vigorous man until he died of cerebral hemorrhage.

Many other well-documented examples could be quoted to demonstrate that multiplication of a virulent microorganism in the body rarely expresses itself in the manifestations of disease. Around 1900 Pettenkoffer in Germany and Metchnikoff in France, with several of their associates, drank tumblerfuls of cultures isolated from fatal cases of cholera. Enormous numbers of cholera vibrios could be recovered from their stools; some of the self-infected experimenters developed mild diarrhea, but the infection did not result in true cholera. More recently human volunteers were made to ingest billions of dysentery bacilli under conditions assumed to be optimal for the establishment of infection. Enteric capsules full of feces obtained directly from acute cases of bacillary dysentery in man were used as additional experimental refinements to increase the chances of establishing the disease. Yet only a few of the volunteers developed symptoms referable to dysentery and most of them remained unaffected by the experimental infection.

The ease and predictability with which Pasteur, Koch, and their followers produced disease at will in experimental animals seem miraculous in view of the difficulties that have so often been encountered in subsequent attempts to produce disease in man. Their success seems incompatible with the course of natural events. The fact of the matter is that Pasteur and Koch did not deal with natural events, but with experimental artifacts. The experimenter does not reproduce nature in the laboratory. He could not if he tried, for the experiment imposes limiting conditions on nature; its aims are to force nature to give answers to questions devised by man. Every answer of nature is therefore more or less influenced by the kind of questions asked.

The art of the experimenter is to create models in which he can observe some properties and activities of a factor in which he happens to be interested. Koch and Pasteur wanted to show that microorganisms could cause certain manifestations of disease. Their genius was to devise experimental situations that lent themselves to an unequivocal illustration of their hypothesis—situations in which it was *sufficient* to bring the host and the parasite together to reproduce the disease. By trial and error, they selected the species of animals, the dose of infectious agent, and the route of inoculation, which permitted the infection to evolve without fail into progressive disease. Guinea pigs always develop tuberculosis if tubercle bacilli are injected into them under the proper conditions; introduction of sufficient rabies virus under the dura of dogs always gives rise to paralytic symptoms. Thus, by the skillful selection of experimental systems, Pasteur, Koch, and their followers succeeded in minimizing in their tests the influence of factors that might have obscured the activity of the infectious agents they wanted to study. This experimental approach has been extremely effective for the discovery of agents of disease and for the study of some of their properties. But it has led by necessity to the neglect, and indeed has often delayed the recognition, of the many other factors that play a part in the causation of disease under conditions prevailing in the natural world—for example, the physiological status of the infected individual and the impact of the environment in which he lives.

Since several distinct determinants usually play a part in the causation of disease processes, it is customary to consider that there are several categories of causes with different levels of importance. Textbooks contrast "initiating," "exciting," or "immediate" causes with "contributory" causes, which play their part merely by bringing the patient under the influence of the initiating causal agent. Simpler and more useful, perhaps, is the recognition of predisposing causes, precipitating causes, perpetuating causes. However, the qualificative appended to a cause is to a large extent a reflection of the present state of knowledge and of prevailing interest. While these differentiations are of help in teaching, they often paralyze thought, and they rarely constitute useful guides for action.

Consider, for example, the evolution of the knowledge of cholera during the past hundred years. John Snow achieved fame for recognizing that an outbreak of cholera

in London affected only persons using the water supplied
by one particular public pump located on Broad Street.
He concluded that cholera was water-borne and he controlled the outbreak by the mere artifice of removing the
handle from the pump. Bad, impure water was for John
Snow the initiating, precipitating cause of cholera. It is now
known that those who used the Broad Street pump contracted cholera because the water that they obtained from
it contained cholera vibrios. As a result, the vibrio is presently considered *the* cause of the disease. But this statement is not so meaningful as appears on first sight since,
as already mentioned, vibrios can be ingested in enormous
numbers and persist in the stools without seriously incon-
veniencing the infected person.

The most that can be said, therefore, is that, once the
vibrios have become established in the intestinal tract, some
other factor can convert the infection into disease. There is
still mystery concerning the circumstances which transform
cholera from the minor nuisance of the bazaar into a raging
pestilence or concerning the factors which bring about the
spontaneous termination of catastrophic outbreaks. But,
while knowledge of the cholera vibrio has not yet proved
very helpful in the understanding of epidemics, much progress has been made in the treatment of the cholera patient.
Effective therapy has followed recognition of the fact that
the most important symptoms of the disease are due to the
loss of fluid and electrolytes from the intestinal tract. Cholera can be successfully treated merely by replacing fluid
and electrolyte, without any serum or antimicrobial drug
to combat the infection. Thus, the mechanisms which increase permeability of the gut might be regarded as the real
cause of the disease since they account for its symptoma-
tology and since treatment of the effects usually results in
cure.

The complexity of most ecological systems renders it difficult to single out any one particular component of the
system as playing a role of unique importance in the causation of disease. Until 1940 all medical textbooks agreed
that the green streptococcus was by far the most frequent

cause of subacute bacterial endocarditis. And indeed it was shown around 1945 that this otherwise fatal disease could often be arrested with doses of penicillin large enough to inhibit the streptococcus. Unfortunately it was soon recognized that disappearance of the streptococcus was not uncommonly followed by infection of the heart valves with other kinds of bacteria normally present in the intestinal or respiratory tract, which penetrate sporadically into the blood stream and settle wherever conditions are favorable for them. Thus, both the initial organic lesions on the heart valve and the various kinds of bacteria that can proliferate on them in succession can be properly regarded as causes of subacute bacterial endocarditis.

The same difficulty is met in trying to determine the cause of deaths during and after episodes of smog. During the winter of 1952 a few days of smog in London resulted in the death of some 5,000 persons, and likewise recent episodes of smog over the Meuse Valley in France or in Donora, Pennsylvania, greatly increased mortality. In general, the deaths that occur during or immediately after a smog are listed in official records as being due to cardiac accidents and bronchitis. On the other hand, there is no doubt that heart disease is extremely widespread and has many different origins, and that bronchitis is associated with the activity of viruses and bacteria ubiquitous in all human communities. What, then, are the causes of deaths that follow smogs? The vascular lesions which are so common in modern man? The bacteria and viruses which almost everyone carries in his respiratory tract? or the poisonous substances in the air which reach everybody but kill only a few?

Direct and Indirect Effects of the External Environment

The process of living involves the interplay and integration of two ecological systems. On the one hand, the individual organism constitutes a community of interdependent parts—cells, body fluids, and tissue structures—each of which is related to the others through a complex network of balance mechanisms. This intra-individual community operates best when its own *internal environment* remains stable within a fairly narrow range characteristic of each species. On the other hand, each organism constantly reacts and competes with all the living and inanimate things with which it comes into contact. Under normal conditions the *external environment* changes constantly, in an unpredictable manner. Many of the modifications that occur in the outer world can have damaging effects. In order to survive and to continue to function effectively the organism must make adaptive responses to these modifications. It must, as well as possible, repair destructive tissue damage and restore its own internal environment to a normal state. Thus, any factor that upsets the equilibrium of either one of these two ecological systems—the internal and external environments—can become a determinant of disease. As all components of both systems are interrelated, any disturbance in either of them—even though minor and not damaging in itself—can set in motion secondary effects which become destructive to the organism Because the process of living necessarily involves all these complex interrelationships, any given pathological process is the resultant of a multiplicity of diverse influences, and all its phases are affected by the adaptive responses to anything that impinges upon the organism.

Ancient physicians knew that the severity and prevalence of various diseases differed greatly according to the geographical area, the time, the social customs, the economic status, the occupation. In the past this dependence was emphasized chiefly with reference to the "fevers," simply because infections like malaria were then so common. It is becoming clear that the environment plays a large part also in determining the prevalence of the diseases most talked about in our times—defects of the cardiovascular system, cancers of various types, peptic ulcers, mental disorders, etc. This is evident from the fact that, as was the case for the fevers in the past, the frequency of the modern diseases differs from one place to another and varies with economic status and professional activities.

There are a few situations in which the damage caused by the external environment results from a direct injury. More often, however, the damage is the indirect outcome of a chain of linked reactions, through which primary impact, which may be innocuous in itself, calls into play deleterious tissue responses. Thus, excessive heat or cold can cause damage directly through destruction of tissue or by lowering or raising body temperature to a level incompatible with life. But it can also elicit reflex contraction or dilatation of the blood vessels, thereby upsetting the circulatory system, with many indirect and possibly fatal consequences. Although there is still much to be learned of the complex physiological processes responsible for the accidents associated with heat prostration and exposure to cold, they are understood sufficiently well to permit some restorative measures. One of the most intriguing applications of this knowledge has been the recent demonstration that it is possible to maintain experimental animals at a freezing temperature, then to thaw them in such a manner that they survive and come back to a normal state.

Like heat or cold, radiations can cause direct damage to the exposed tissues and bring about their death. But radiations also have indirect effects probably more important under the ordinary circumstances of life. In animals, and probably in man also, they facilitate the passage of the microorganisms from the intestine into the general circulation and thereby indirectly cause disease through the toxicity of infection. The effects of ionizing radiations in man have been highlighted recently by the reports of commissions appointed for their study in America and England. It had long been known that certain skin cancers occur more frequently in people much exposed to sunlight. Evidence is now accumulating that the incidence of leukemia is increased in patients receiving radiotherapy and that radiologists have a shorter expectancy of life than other physicians of the same age group not similarly exposed. In Japan an increase in premature senility and in early death has been noted among persons who were in the vicinity of the atomic explosions. Indeed, there is evidence that exposure to radiation constitutes a potential danger even in less dramatic situations—for example, during the manufacture of luminescent watch dials or in the course of medical examinations with X rays for diagnostic work.

Least well understood, but perhaps most important for mankind, is the influence that radiations may exert on future generations by their long-term genetic effects. Throughout his life and evolution man has been exposed to a background of radiation from natural sources. This normal background, which seems to have remained fairly constant, at least in geologically recent times, must therefore be a tolerable factor of human environment. But any increase in dose rate of radiation is likely to increase the rate of mutations—with unforeseeable consequences. And it is unlikely that adaptive processes can occur fast enough to cope with the potential long-range dangers if man-made radiation continues to increase at its present rate. The atomic age will probably bring in its train many obscure pathological disorders even if effective steps are taken to avoid obvious radiation burns

Among other factors of the environment that have direct effects in the production of disease are certain toxic substances ingested in the food. For example, in India and other places where chick-peas constitutes a large percentage of the diet, lathyrism is a very common affliction whenever there is a shortage of other food products. Lathyrism is caused by a constituent of certain legumes which is toxic for nervous tissues. Many diseases of farm animals have been similarly traced to toxic substances present in forage—the hemorrhagic agent coumarin in fermented sweet clover, the steroid produced by subterranean clover which causes abortion in cattle Nutritional deficiencies, especially of vitamins, proteins, or minerals, are so well known in man and animals that they need no further emphasis. But in addition to these fairly direct effects of faulty nutrition there are others more indirect in their determinism—for example, the shortening of life expectancy correlated with too rich and abundant nutrition. The deficiency in vitamin B_{12} in sheep fed on pastures which do not provide the proper amount of cobalt for their intestinal bacteria is another example of indirect deleterious effect of the environment (See Chapter III).

The role of the external environment in microbial diseases has been considered chiefly with regard to transmission of the infectious agents. Many of the examples relating the occurrence of "fevers" to certain locations, as mentioned so frequently in Hippocrates' classic treatise on "Air, Water and Places," can be traced to the presence of the mosquitoes transmitting malaria. Clearly much of the effect of environment on the frequency of microbial diseases is thus a direct expression of the manner in which viruses, bacteria, fungi, or higher parasites are transmitted through the air, the water, the food, or insect vectors. But, important as it is, this mechanism does not account for all the effects of the external environment on microbial diseases.

The course of psittacosis in parakeets, for example, is influenced profoundly by factors of the environment which do not act directly on the virus or affect its transmission. In general the young birds are infected with the psittacosis virus while still in the nest. The infection remains inapparent in most of the birds, manifesting itself only by the intermittent release of virus in the droppings. But many factors can upset the equilibrium between virus and infected bird. Crowded and unsanitary conditions of husbandry, shipping to distant markets, intensive breeding—all these and other ill-defined circumstances are liable to result in unrestrained multiplication of the virus, with overt disease and death of the birds. In a large aviary of two thousand parakeets in which breeding was stopped by separating the sexes, deaths from psittacosis disappeared entirely within two months However, the infection was not eliminated thereby. It had merely become silent, and it flared up in an active form again within five weeks after mating and breeding was resumed.

The causation of many disease processes in man bears a close analogy to psittacosis in birds—for example, the familiar fever blisters produced by the herpes simplex virus. The herpes virus is usually acquired during childhood. Throughout most of the life of the infected individual it lies latent in the body, without causing any symptom or obvious pathology until "provoked" into activity by some physiological disturbance. As is well known, herpetic fever blisters can be elicited by a variety of unrelated stimuli, ranging from colds and fevers of various origin to menstruation, ultraviolet radiation, or eating cheese. Herpetic blisters thus provide a striking example of an infectious disease of man in which, contrary to the original tenets of the germ theory, the living agent of the disease can be present all the time in the host—be intrinsic, so to speak—whereas the provocative determinant of the pathological process is some physiological disturbance or some other extrinsic factor of the environment.

Similarly, the most important problems of infection among internees in the German concentration camps during World War II were not the acute epidemics of exotic nature, such as cholera or typhus. Rather, they were the commonplace colds, bronchopneumonias, skin infections, pulmonary tuberculosis, etc., all conditions caused by organisms normally endemic in European communities. While life under normal conditions brings about in most individuals a satisfactory *modus vivendi* with these ubiquitous pathogens, malnutrition and other forms of physiological misery caused infection to evolve into overt disease in many internees It is of special interest in this regard that after returning to their normal environment at the end of the war most internees rapidly recovered from the microbial diseases without the help of specific therapy.

Psittacosis, herpes blisters, all sorts of ill-defined respiratory disorders, are typical of a host of ecological situations which account for the largest bulk of microbial diseases in nature. In these situations the microbial agent is ubiquitous in the community and can persist in the tissues of the individual without causing detectable damage, but infection can be converted into overt disease by all sorts of disturbances resulting in physiological or mental misery.

A case recently brought to trial in an English court of justice illustrates how the complex character of ordinary ecological situations often creates unexpected difficulties in establishing a legal *cause* for a banal disease. According to an account published in *The Lancet* of November 1954, a lacquer sprayer sued his employers on the ground that he had contracted pneumonia and pleurisy because the room in which he had worked was cold and drafty. His Lordship the judge found that the plaintiff's working place was indeed cold, drafty, and damp in the early morning. He accordingly awarded damages totaling £ 401, feeling satisfied that the plaintiff's illness was *caused* by the absence of heating There is little doubt that the pneumonia and pleurisy of which the lacquer sprayer complained were manifestations of the activities of some microbial agent— virus or bacterium or probably both. Furthermore, it is likely that the workman had not contracted infection in the shop but had been harboring the guilty microbes in his organs for weeks, months, or probably even years. In ruling that the deficient heating had *caused* the pneumonia, his Lordship has assumed—wisely, it would seem—that the environment is often as important as the microbe in the determinism of microbial disease.

The Internal Environment

Despite its conceptual simplicity, almost pedestrian in obviousness, the doctrine of specific etiology played little part in the development of medical thought before the Pasteur-Koch era. Until the late nineteenth century health had been regarded as harmony between the individual and his environment and between the various forces at work in the body Disease occurred when the equilibrium was disturbed, by any cause whatever. This medical philosophy has been expressed in many different forms throughout history. Among primitive people it echoes the Navajo wish to live "in accord with the mountain soil, the pollen of all the plants, and other sacred things." To maintain a proper balance between the yin and the yang was the goal of the Chinese physician, just as the proper balance among the four humors of the body was the most important aspect of Hippocratic teaching. Rudolf Virchow regarded disease as "life under changed conditions" and concluded therefrom that the physician had to concern himself with the total environment of human beings and therefore could not avoid taking part in political action. In a still different form the same doctrine came to light in the heated controversies stimulated by Pasteur's interventions in the French Academy of Medicine during the 1880's. Disease, asserted his opponent Pidoux, "is the common result of a variety of diverse external and internal causes . . . bringing about the destruction of an organ by a number of roads which the hygienist and the physician must endeavour to close."

By equating disease with the effect of a precise cause—microbial invader, biochemical lesion, or mental stress—the doctrine of specific etiology had appeared to negate the philosophical view of health as equilibrium and to render obsolete the traditional art of medicine. Oddly enough, however, the vague and abstract concepts symbolized by the Hippocratic doctrine of harmony are now re-entering the scientific arena. Hippocratic medicine has acquired a more profound significance from the implications of the discoveries that Darwin and Claude Bernard were making around 1850—even before Pasteur and Koch had made their contributions to the etiology of disease. Darwinism implies that the individual and species which survive and multiply selectively are those best adapted to the external environment. Claude Bernard supplemented the doctrine of evolutionary adaptation by his visionary guess that fitness depends upon a constant interplay between the internal and the external environment of the individual. He emphasized that at all levels of biological organization, in plants as well as in animals, survival and fitness are conditioned by the ability of the organism to resist the impact of the outside world and maintain constant within narrow limits the physicochemical characteristics of its internal environment. In other words, life depends not only upon the re-

actions through which the individual manages to grow and
to reproduce itself but also upon the operation of the control mechanisms which permit the maintenance of individuality. *"La fixité du milieu intérieur,"* Claude Bernard
wrote, *"est la condition essentielle de la vie libre."* It is
indeed the very condition of conscious life.

The dual concept of fitness to the external environment
and fixity of the internal environment is the modern expression of the Hippocratic dictum that health is universal sympathy. Like all fruitful concepts, the views formulated by
Darwin and Claude Bernard have continued to evolve, becoming more precise in detail and broader in outlook. The
science of biochemistry has provided exquisite evidence of
the fact that the composition of the body fluids and tissues
remains constant within narrow limits and has unraveled
the chemical reactions through which this dynamic con-
stancy is achieved. The science of physiology has revealed
that the wisdom of the body, its ability to respond adaptively to all sorts of situations, depends upon the mobiliza-
tion and complex interplay of the various hormones which
govern homeostatic mechanisms. Homeostasis is evident
also in pathological situations, the body attempting—usually
with a certain measure of success—to protect itself against
injury through inflammatory and repair processes. The
theory of the general adaptation syndrome is the most recent conceptual scheme devised to embrace all the various
mechanisms through which the body responds to the "stress
of life." Following Freud, psychiatry is formulating its problem in the same light. Like physiological responses, emotional reactions often constitute unconscious mechanisms—
instinctive or acquired—through which the organism attempts to defend itself against threats, real or symbolic.

During recent years it has become apparent that all homeostatic mechanisms are linked and operate jointly in the
normal organism to meet the immense variety of stimuli
and threats which are part and parcel of life. These defense
mechanisms, however, are not merely passive. Their operations involve on the part of the body powerful reactions
which tend to ward off the threat or to repair the damage

done In this respect the concept of homeostasis is analogous to the Hippocratic view that disease involves not only suffering (*pathos*) but also work (*ponos*). *Ponos* is the work expended by the body in its attempts to achieve homeostasis and maintain its individuality in an ever-changing world. It is true, of course, that there are only a limited number of ways in which the organism can respond to stimuli and that the response is not always beneficial. It may be too weak, too strong, or misdirected. As the manifestations of this faulty response are limited in number and character, disease, which is made up of their summation and of their consequences, can take only those forms of which the organism is capable, and cannot possibly reflect the peculiarities of each of the stimuli from which it originated.

In the light of these facts it is easier to understand why direct cause-effect relationships often fail to account for the natural phenomena of disease. Each type of insult—microbial invasion, chemical damage, physiological stimulus, or psychic event—can have many different effects depending upon the state of the recipient individual. On the other hand, any given pathological effect can be the outcome of many varied kinds of insults. Under well-defined conditions each type of insult will of course bring about predominantly one type of pathological effect. The approach of the experimenter is to take advantage of this fact by devising laboratory models which produce the desired phenomena. This permits him to study in detail the mechanisms through which the various factors exert their effects and the manner in which these can be modified by willful and rational intervention.

In contrast to the experimenter, the epidemiologist has to deal with biological phenomena in all their natural complexity. He must try to recognize the relationships most common in certain specified ecological situations and to derive from this knowledge the methods of control having the best statistical chance of being useful in each particular situation. Whereas the experimenter determines by choice the phenomena that he studies and the epidemiologist formulates statistically those most significant under the conditions that he observes, it is the responsibility of the physician to decipher the relative importance of the various factors involved in the response of each individual patient and to decide which aspects of the internal and external environment can be safely manipulated for the purpose of treatment. It is because each clinical decision involves so many judgments of facts and of values that medicine in its highest form will continue to remain an art.

Interplays between the External and the Internal Environment

Any event in the outer world which impinges on an individual modifies, however indirectly and slightly, the balance between his various organs and functions. In reality, therefore, the internal environment should not be considered apart from the external environment. Shivering or pallor, brought about either by exposure to cold or by a sudden fear, is but the outward manifestation of a physiological disturbance which may alter indirectly the performance of many essential body mechanisms. The temporary contraction of blood vessels revealed by pallor may increase resistance to blood flow, and this change in the state of the vascular bed may overload the heart. Pallor can also be seen to occur on the inner surface of the stomach in certain emotional states, thus making clear that the whole digestive tract responds to emotions in a manner not unlike that of the rest of the body. On the other hand, contraction of the blood vessels necessarily interferes with the nutrition of the affected part of the body both by limiting the amount of blood available to it and by interfering with the removal of the waste products of metabolism. There is little doubt that these physiological disturbances often bring about lesions resulting from local changes in tissue cells and fluids or from the activities of microbial agents which would otherwise be held in check by the normal tissue response.

All important pathological disorders are the summation of a multiplicity of interplays between the external and the internal environment. The factors involved in the genesis of mammary cancer in mice, and of certain plant cancers, provide illuminating examples of these complex interrelationships.

Mammary carcinoma occurs as it were spontaneously in a very large percentage of breeding females in certain inbred strains of mice. However, production of the carcinoma depends upon the presence of a virus-like agent which in nature is transmitted through breast feeding, as shown by the fact that foster nursing of newborn mice by mothers from a strain with low tumor incidence renders the latter animals free of the tumor tendency. Thus, it might be concluded that the "milk virus" is the cause of the carcinoma. In reality the situation is far more complex than would appear from these facts. In inbred strains of mice with a high tumor rate the milk virus is present in all animals of the colony, males as well as females, even though it may fail to produce any tumor or even any sign of disease. Male mice do not develop the tumor, nor do females until under the stress of repeated breeding and lactation. On the other hand, injection of estrogenic hormone will induce the tumor in males of inbred strains in which a high incidence occurs in lactating females but not in females of a strain with low incidence. In brief, three factors at least are involved in the genesis of mammary cancer in inbred mice: the milk virus, a hormonal influence, and a genetic susceptibility to the action of the first two. Moreover, the incidence of mammary cancer is drastically reduced when mice with all these known factors are placed on a limited calorie intake. It is clear, therefore, that several physiological factors play a part as important as that of the virus in the causation of mammary carcinoma in mice.

A situation of analogous complexity obtains with regard to the tumors (crown galls) commonly seen on sunflowers and certain other plants. Under the proper conditions the tumor can be produced at will by injecting into the plant a certain kind of bacterial culture (*Agrobacterium tumefaciens*). Under other conditions, however, infection with these bacteria results not in the production of tumor but in extensive invasion of the plant. Only plant tissues which have been conditioned by certain stimuli associated with wound healing can produce tumors under the stimulus of infection. Intensive study of this problem has revealed many other phenomena—not to be discussed here—which implicate complex genetic, biochemical, and nutritional determinants in the causation of the crown gall. Thus, it is clear that in plants as well as in animals several independent factors must operate simultaneously in order to give rise to disorganized growths. Like other forms of disease, most cancers are probably the resultant of a constellation of circumstances. This complexity appears disheartening to the investigator, but it is also a source of hope, since the multiplicity of links in the chain of causation increases the chance of finding a weak part in the chain and thus of discovering a method of control.

The Past as Factor of the Environment

In most cases, the effect produced by any stimulus is conditioned by the biological and social history of the group and by the past experience of each individual. In other words the manner of response is predetermined not only by the molding effect of selective evolutionary forces but also by the accidents of personal life—from allergic idiosyncrasies to acquired patterns of behavior. In man a particular odor may be pleasurable or disgusting, create an atmosphere of hope or of despondency, depending upon the mood of the day in which it was first experienced and which it evokes out of the biological memory. One man's meat is another man's poison, not only by reason of genetic differences but also because some of the physiological and psychic reactions to food are modified or determined by past individual experiences, most of them unnoticed at the time of their occurrence or long since forgotten. It is because the effects of early influences persist in the tissues in such a reactive state that the past is ever with us. The effects of the physical and social environment cannot be understood without knowledge of individual history.

Little is known of the forms in which biological memory persists in the tissues or of the mechanisms that make the past exert effects peculiar to each individual. In fact, it is only in the case of immunologic and allergic phenomena and also of some psychic reactions that techniques have been developed to bring out experimentally the manifestations of the past in the living organism.

The word "allergy," so commonly used today, is of recent origin. It was coined from the Greek some five decades ago to denote a change in the response of the body to a given substance. The individual who exhibits allergy to a substance must have been exposed to it under the proper circumstances at some time in his past. Thus, Europeans as a group do not exhibit allergic sensitiveness to poison ivy, simply because this plant does not exist in Europe, but they become allergic to it just as readily as do Americans after repeated exposure.

Allergy is only one of the countless manners in which past experience conditions the response of the individual, and conditioning can occur not only in man and highly evolved organisms but also in the most primitive living things. At the turn of the century, it was a popular exercise among biologists to devise experiments illustrating that exposure to a given stimulus or substance modified the response of even the most lowly organisms—bacteria, algae, protozoa—to subsequent contact with the same stimulus or substance. These striking changes in response are no longer fashionable subjects of investigation, probably because they cannot as yet be analyzed by physicochemical methods. Still they remain of interest in showing that manifestations of behavior at primitive levels of organization present analogies to those encountered in human life—as appears from the following observations made on sea urchins.

Because light seems to be somewhat injurious to it, the sea urchin naturally tends to remain in dark places. Yet it responds to a sudden shadow falling upon its body by pointing its spine in the direction from which the shadow comes. This response is defensive, serving to protect the sea urchin from enemies that might have cast the shadow in approaching. The reaction is elicited by the shadow, but it refers to something symbolized by the shadow. As is well known, similar symbolic reactions reach a complex development in higher animals. In man, indeed, practically all responses to things seen or heard are merely reactions to representative stimuli. Just as comparative studies of metabolism have revealed a remarkable unity in all the biochemical phenomena of life, so studies of behavior have brought out certain patterns which, formally at least, are common to all living things.

The response to symbols of stimuli calls to mind, of course, the conditioned reflexes made popular by the Russian physiologist Pavlov. It has a bearing also on the lasting influence that almost any event of the past exerts on the behavior of normal human beings. In every individual certain life situations or sensations call forth almost compulsive responses arising from association with some event usually forgotten. Ever since Marcel Proust, the reading public has become almost pathologically aware of the enormous part played in emotional life by the biological Temps Perdu. A madeleine dipped in a cup of tea, or the uneven cobblestones in a Paris courtyard, can reactivate the serenity or the heart pangs of youth with all their original intensity and purity. Less well recognized is the fact that Things Past affect profoundly every aspect of organic life. A few decades ago the American behaviorists succeeded in conditioning the response of a child to a white rabbit. More recently Russian physiologists have shown that the use of a certain numeral during an ordinary exercise in arithmetic always brings about excessive salivation in children who have been conditioned to that numeral sometime before the exercise.

While salivation is a response that can be seen, there are other responses, far more significant, that are not so readily detected. We are aware of the sense of tension, the quickened pulse, the flush or pallor, a moment of perspiration, but these are only outward manifestations of more fundamental disturbances. What we cannot evaluate is the cumulative effects of these disturbances on the vascular bed, as well as on other essential structures and functions. Clearly the increased salivation of the child in response to a certain numeral and the tenseness evoked by the sound of a voice are but the representative signs of all the inner reactions that produce lasting scars in our physical and emotional lives.

Thus, the internal environment is constantly responding to the external environment, and history—racial, social, as well as individual—conditions the manner of response just as much as does the intrinsic nature of the stimulus. Moreover, past experiences render the individual susceptible even to symbols of the stimulus. Man, like the sea urchin, responds in a compulsive manner not only to actual threats or to the presence of enemies but also, and even more strongly at times, to the many shadows which have come by association to symbolize danger. Environment cannot be defined merely in terms of the intrinsic characteristics of the microbes, poisons, radiations, trauma, and stresses to which the individual is exposed at a given time. It must include also the phenomena which have affected him in the past. Man can truly say, like the old Greek warrior in Tennyson's poem, "I am a part of all that I have met."

V.

HYGEIA AND ASCLEPIUS

Gods of Health

The word "hygiene" now conjures up smells of chlorine and
phenol, pasteurized foodstuffs and beverages in cellophane
wrappers, a way of life in which the search for pleasurable
sensations must yield to practices that are assumed to be
sanitary. Its etymology, however, bears no relation to this
pedestrian concept. Hygiene is the modern ersatz for the
cult of Hygeia, the lovely goddess who once watched over
the health of Athens Hygeia was probably an emanation,
a personification of Athena, the goddess of reason. Although
identified with health, she was not involved in the treatment of the sick. Rather, she was the guardian of health
and symbolized the belief that men could remain well if
they lived according to reason. There is in the Museum of
Athens a lovely marble head, probably originating in Tegea
in Arcadia about 380 B.C. It shows Hygeia as a serene,
benevolent maiden, personifying health by her balanced
and reasonable demeanor.

Throughout the classical world Hygeia continued to
symbolize the virtues of a sane life in a pleasant environment, the ideal of *mens sana in corpore sano*. In Greece she
eventually came to be more closely identified with mental
health and in Rome she was known as Salus, a divinity of
well-being in general. But in reality Hygeia was not an
earthbound goddess of ancient origin. Her name derives
from an abstract word meaning health. For the Greeks she
was a concept rather than a historical person remembered

from the myths of their past. She was not a compelling
Jeanne d'Arc but only an allegorical goddess Liberty and
she never truly touched the hearts of the people. From the
fifth century B.C. on, her cult progressively gave way to that
of the healing god, Asclepius.

To ward off disease or recover health, men as a rule find
it easier to depend on healers than to attempt the more
difficult task of living wisely. Asclepius, the first physician
according to the Greek legend, achieved fame not by teaching wisdom but by mastering the use of the knife and the
knowledge of the curative virtues of plants. In contrast to
Hygeia, the name Asclepius is of very ancient origin. Apparently Asclepius lived as a physician around the twelfth
century B.C. He was already known as a hero during
Homeric times and was created a god in Epidaurus around
the fifth or sixth century B.C. His popularity spread far and
wide, even beyond the boundaries of Greece. Soon Hygeia
was relegated to the role of a member of his retinue, usually
as his daughter, sometimes as his sister or wife, but always
subservient to him. In most of the ancient iconography from
the third century on, as well as in all subsequent representa-
tions, Asclepius appears as a handsome, self-assured young
god, accompanied by two maidens: on his right Hygeia and
on his left Panakeia. Unlike Hygeia, her sister, Panakeia became omnipotent as a healing goddess through knowledge
of drugs either from plants or from the earth. Her cult is
alive today in the universal search for a panacea.

The myths of Hygeia and Asclepius symbolize the never-
ending oscillation between two different points of view in
medicine. For the worshipers of Hygeia, health is the natural order of things, a positive attribute to which men are
entitled if they govern their lives wisely. According to them,
the most important function of medicine is to discover and
teach the natural laws which will ensure to man a healthy
mind in a healthy body. More skeptical or wiser in the ways
of the world, the followers of Asclepius believe that the
chief role of the physician is to treat disease, to restore
health by correcting any imperfection caused by the accidents of birth or of life.

In one form or another these two complementary aspects
of medicine have always existed simultaneously in all civilizations. It is written in the Yellow Emperor's *Classic of
Internal Medicine* that "the ancients followed Tao and the
laws of the seasons under the guidance of their sages who
were credited with the realization of the value of education
in the prevention of disease." Whatever the precise meaning
of this statement, it certainly implies that ancient Chinese
medicine embodied rules of behavior thought essential to
the maintenance of health. But Chinese medicine also developed a very evolved system of therapeutics with complex
surgical procedures and many useful drugs. Even its method
of acupuncture, long a subject of skeptical scorn on the part
of Western physicians, is now believed to have some valid
physiological basis.

A similar coexistence of hygienic wisdom in the affairs
of daily life and of valuable surgical and medical lore is
found among most primitive peoples. Taboos, religious
practices, and ancestral customs which appear meaningless
to us had in many cases a definite health function at the
time they were formulated. The great leaders of men in
the past—the Moses and Mohammed of the ancient world—
owed part of their success to the sanitary discipline that
they enforced in nomadic life. As is beginning to be realized,
there is more to environmental sanitation than the technical plumbing of Westernized cities. Hygiene involves a social philosophy that must take into consideration the human and economic aspects of the cultural pattern for which
it is designed. Just as the cult of Hygeia idealized the
Greek way of life, many rules of conduct represent the attempts of peoples to achieve an adaptation to their environment compatible with physiological needs and emotional urges.

Needless to say, this adaptive state often breaks down,
and for many reasons. Men will rebel and transgress, un-
deterred by the consequences of their actions for themselves
and their group. Furthermore, empirical rules of conduct
lose their usefulness when applied under biological or social
conditions different from those under which they had

evolved. They are solutions reached by trial and error, to meet special circumstances, but are rarely applicable to other conditions. No habit or rule of conduct, whatever its wisdom, can take care of antisocial behavior or of unforeseeable accidents. As disease and other failures of adaptation are obvious and often dramatic, whereas health and fitness are considered the "normal" state and therefore unnoticed, it is not surprising that the cult of Hygeia tends to be neglected and that the skill of Asclepius looms large and bright in the mind of man. In our societies the school of public health always plays second fiddle to the school of medicine.

Most civilizations very early developed a class of specialists possessing great therapeutic skill. They practiced massage, bloodletting, dry cupping, cauterization; they performed astonishing surgical feats, from the treatment of fractures to removal of kidney stones and to trephining. Some of the techniques are very remarkable indeed; for example, the suturing of wounds with termites as practiced in India, East Africa, and Brazil. In brief, the procedure consists in bringing the edges of the wound close together, having termites bite through them, whereupon the heads are cut off! Primitive people have also recognized the virtues of many effective drugs, most of which are still in use today, and have applied them wisely to the relief of human suffering. Opium, hashish, coca, cinchona, podophyllin, and ephedrine are but a few of the natural products that have come to us as a heritage from primitive medicine. That more rediscoveries can be expected is shown by the fact that extracts of the plant Rauwolfia, a drug long known to Hindu physicians, have recently found an important place in Western medicine as tranquilizer and for the treatment of certain mental diseases. Thus, an immense amount of useful knowledge of therapeutics had accumulated long before the so-called scientific era. Indeed, it can be said that the sciences of anatomy, physiology, and in part of chemistry found their raw material in the discoveries of the disciples of Asclepius and of his counterparts all over the world. In contrast, it is not apparent that the cult of Hygeia contributed much to scientific development, unless a life of reason be considered a prerequisite to science. While Asclepius is in Luther's words only "God's body patcher," the serene loveliness of Hygeia in the Greek marble symbolizes man's lofty hope that he can someday achieve a state of harmony within himself and with the surrounding world.

Hippocratic Wisdom and the Gold-headed Cane

Hippocratic writings occupy a place in medicine corresponding to that of the Bible in the literature and ethics of Western peoples. Just as everyone quotes from the Bible, it is the universal practice to look to Hippocrates for statements that give the sanction of authority and of time to almost any kind of medical views, profound or banal. But very few of those who quote Hippocrates have ever read him, even though his works have often been translated and annotated. The immense and lasting prestige of his writings is due in part to the many-faceted aspects of their message. As in the Bible, again, everyone can find in them something relevant to his preoccupations which has never been stated better and more succinctly.

To the philosophically minded, Hippocrates stands for rational concepts based on objective knowledge and for the liberation of science in general, and of medicine in particular, from mystic and demonic influences. To the student of public health, the Hippocratic writings provide classical examples of the relation between the environments—physical and social—and the prevalence and severity of various diseases. To the physician, they offer objective descriptions of symptoms, subtle methods of diagnosis, hints for dealing with the patient as a person and with his family, in short, all aspects of the art of medicine. For twenty-five centuries Hippocrates has personified in the Western world the rational outlook of the philosopher, the objective attitude of the scientist, the practical approach of Asclepius, and the human traditions of Hygeia.

Because Hippocrates has become everything to all men, it is probable that we read into his utterances meanings that are clearer, more profound, and more universal than those recognized by the ancient world. This is, of course, true of all great works of philosophy, literature, art, and science, each generation emphasizing those aspects of them which are peculiarly relevant to its own problems. The very concepts of health and disease have in the course of time evolved progressively as a flowering of Hippocratic influence.

In most primitive cultures disease is regarded as a visitation of some hostile god or of other capricious forces. The Jewish tribes believed that obedience to Jehovah's laws was a necessary passport to good health and that any transgression was likely to be punished by disease. Illness was the wage of sin. "If thou wilt diligently hearken to the voice of the Lord . . . I will put none of these diseases upon thee." Moreover, God was the supreme healer. In contrast, it is implicit in the Hippocratic teachings that both health and disease are under the control of natural laws and reflect the influence exerted by the environment and the way of life. Accordingly, health depends upon a state of equilibrium among the various internal factors which govern the operations of the body and the mind; this equilibrium in turn is reached only when man lives in harmony with his external environment.

Ancient medicine never achieved an understanding of the forces controlling the internal environment; indeed, this knowledge began to develop barely a century ago. But one finds in the Hippocratic writings many details concerning the influence that the external environment exerts on human affairs. Most often quoted are the passages in the books on "Regimens" and on "Air, Water and Places," which deal with the prevalence and severity of organic diseases. Hippocrates' interests, however, extended also to the effect of the environment on the other aspects of human life. Rough countries with excessive seasonal changes, he claimed, are apt to endow the men who inhabit them with strength and violent passions, and they tend thereby to interfere with

the peace and gentleness of life. Even the political regime
of the Asiatics and the Greeks he believed of importance
for their physical vigor. Because the Asiatics lived under
despotic rule they were less vigorous and less effective in
war than the Greeks, also living in Asia but enjoying free
democratic institutions.

Hippocrates regarded disease as the result of infringe-
ment upon the natural laws, but the scant knowledge of his
time did not permit any analysis of the mechanisms involved. Nevertheless, his writings are pervaded with the
concept that the life of the patient as a whole is implicated
in the disease process and that the cause is to be found in
a concatenation of circumstances rather than in the simple
direct effect of some external agency. "After Silenus had
helped himself to some cheese before bedtime, he awoke
the next day with painful indigestion. But Proclops, his
friend, who had eaten an equal portion suffered no ill
effects." Since the cheese had affected each man in a different way, Hippocrates reasoned that Silenus, the man,
was sick, and not Silenus' stomach. As it happened, Silenus
had spent a strenuous day at the gymnasium. He was tired
and overheated when he ate the cheese. Thus, a surplus of
"fire" had upset his humoral balance. To correct this, Hippocrates advised Silenus to avoid strenuous exercises or to
cool off before eating.

Fitness to the total environment being essential for health
in the Hippocratic view, it follows that disease almost inevitably ensues when changes in conditions are too rapid
and too violent to allow adaptive mechanisms to come
into play. Time and time again Hippocrates comes back to
the dangers entailed in sudden changes of any sort. "It is
changes that are chiefly responsible for diseases, especially
the greatest changes, the violent alterations both in the seasons and in other things. But seasons which come on gradu-
ally are the safest, as are gradual changes of regimen and
temperature, and gradual changes from one period of life
to another." On the other hand, Hippocrates taught that
anyone living reasonably in an environment to which he was
well adapted was not likely to fall ill, unless some accident

or epidemic occurred. "When an epidemic of one disease is prevalent, it is plain that the cause is not the regimen but what we breathe, and this is charged with some unhealthy exhalation. During this period . . . patients should not change their regimen, as it is not the cause of their disease. For if the change of regimen be sudden, there is a risk that from the change too, some disturbance will take place in the body."

Hippocrates also believed that the sick body calls into play natural forces that tend to restore the disturbed equilibrium and to re-establish health. Consequently the physician was to take advantage of this natural tendency to healing—the *vis medicae naturae*—by providing an environment and a way of life congenial to the patient's nature, with the help of the proper nutritional regimen and of a few palliative drugs. The very word "physician"—from the Greek root meaning nature—was used by Hippocrates to denote that every practitioner of medicine "was to be skilled in Nature and must strive to know what man is in relation to food, drink, occupation, and which effect each of these has on the other." Moreover, the physician should never forget that disturbances in any organ correspond to a disturbance of the whole person and that "to heal even an eye, one must heal the head and even the whole body."

This faith in the healing power of nature was not shared by all physicians of the ancient world. Many probably believed, as did later Asclepiades in Rome, that the expectant attitude of the Hippocratic school was little more than a "mediation on death" and that disease demanded more aggressive treatment. Nevertheless, there is no doubt that Hippocrates' simple therapeutic policies achieved for him a great reputation as a healer, at least if we believe the testimonies of Plato and Aristotle, as well as the words carved on his tombstone in Cos:

HERE LIES HIPPOCRATES
WHO WON INNUMERABLE VICTORIES
OVER DISEASE
WITH THE WEAPONS OF HYGEIA

The immense reputation of Hippocrates as a healer poses
a problem which applies also to all the successful physicians
of the past. Although most of medical science is of very
recent acquisition, the healing act has been practiced with
success for thousands of years without the benefit of the
factual knowledge now considered essential even in the lowest grade medical school Many physicians of the prescientific era achieved great fame through their skill in formulating prognoses, relieving symptoms, and not uncommonly
effecting cures.

It is not easy for the modern scientific doctor to visualize
without a smile his counterpart of not so long ago. The
fashionable English physician of the late seventeenth century went on his sick calls dressed in a silk coat, breeches,
and stockings, with buckled shoes, lace ruffles, full-bot-
tomed wig, and carrying a gold-headed cane. Typical of
this class was John Radcliffe (1650–1714), physician to
William and Mary and to Queen Anne. Radcliffe was apparently a somewhat coarse, pompous, uncouth, and intemperate man with a sharp tongue, who attributed his
success to his habit of "using all mankind ill " But even his
personal enemies acknowledged him to be a sagacious
physician possessed of great powers of observation, medical
acumen, and knowledge of mankind. There was in his ante-
room a bowl into which the patient could drop whatever
gratuities he felt moved to leave. Many a purse of golden
guineas must have been dropped therein by the grateful
recipients of Radcliffe's medical advice; although the fee
was not stipulated, he amassed one of the great fortunes of
medicine which allowed him to endow institutions of learning in London and Oxford. When he retired from practice
in 1713, Radcliffe presented his gold-headed cane to Richard Mead, from whom it passed on successively to Anthony
Askew, William Pitcairn, David Pitcairn, and Matthew
Baillie (1761–1823) Thus, for one hundred fifty years,
Radcliffe's gold-headed cane remained in the possession of
some of the most successful practitioners of England. In
Matthew Baillie, however, it found a master of a different
mentality. Baillie was the nephew of the famous brothers,

John and William Hunter, from whom he had acquired the belief that experimental science had become essential to the advance of medicine and would render obsolete the assumed gravity and pompous airs of fashionable physicians. No longer regarding the gold-headed cane as a necessary appendage of the medical profession, Baillie discontinued carrying it on his sick calls. After his death it was found in a corner of the consulting room and his wife presented it in 1825 to the museum of the Royal Academy of Physicians, where it has remained ever since.

It goes without saying that, despite Baillie's skepticism, many physicians have continued to exert with skill and success a type of care of the sick based more on intuition and art than on modern science. Indeed, many of the most brilliant practitioners of clinical medicine of the nineteenth century held that their art had nothing of importance to learn from biological sciences. Some of them achieved a place in history by the passion with which they opposed the germ theory of disease. The French clinician Pidoux, a representative of traditional medicine—always impeccably clad in a gold-buttoned coat—was one of those who took the cudgels against Pasteur at the Paris Academy of Medicine. Where Pasteur saw disease as caused by specific kinds of microbes, Pidoux invoked the concept of diathesis, emphasizing that any disease could be caused by a multiplicity of internal and external causes and could not be regarded as due to one single agent. Pidoux and his like spoke in broad clinical terms, basing their doctrines on the phenomena of disease as observed in man, but their vague arguments were no match for the precise experimentation by which Pasteur, Koch, and their followers defended the doctrine of specific causation of disease. Experimental science triumphed over the clinical art, and within a decade the theory of specific etiology of disease was all but universally accepted, soon becoming, as we have seen, the dominant force in medicine.

There is little doubt, however, that the successful physicians of the past had learned through practical experience to manipulate empirically many of the factors—physiological, psychological, and social—which affected the reactions of their patients. Helping the patient to rest was certainly part of this poorly defined skill and implied much subconscious knowledge of human nature and customs. Despite the apparent simplicity of the concept, rest in reality involves complex factors colored by habits and emotions, and the manner of practicing it has changed greatly with time. Horseback riding, which would prove today a difficult ordeal for most patients, was much in favor as a form of rest during the eighteenth and the early nineteenth centuries, a time when it constituted a common and often the easiest form of transportation. In his book on consumption published in the seventeenth century, Benjamin Marten expressed the opinion that, whereas prolonged residence in the dark, smelly, and stuffy bedrooms of his time was conducive to physical and mental fretfulness, "riding on Horseback in the manner of Travellers and not furiously" helped to keep the patients entertained in a restful manner.

Ancient physicians, of course, also used their knowledge of human nature to practice some form of psychotherapy; they certainly helped the operations of all the natural forces of resistance to disease by instilling confidence in their patients. As Burton wrote in his *Anatomy of Melancholy*, it is by virtue of the healing power of confidence that "An empirick oftentimes, and a silly chirugeon, doth more strange cures than a rational physician . . . because the patient puts his confidence in him, which Avicenna prefers before art, precepts, and all remedies whatever. . . . He doth the best cures, according to Hippocrates, in whom most trust."

For all these reasons, the skill symbolized by the goldheaded cane was not mere charlatanism. It grew in no small part from the physician's awareness—even though ill defined and often subconscious—of the many factors which play a part in the causation and manifestations of disease. As recently as half a century ago Osler was wont to assert that "It is much more important to know what sort of a patient has a disease, than what sort of disease a patient

has!" The art of medicine was one of the fruits of the Hippocratic flowering.

The Philosopher's Search for Health

Hippocrates taught, as we have seen, that man would
have a good chance of escaping disease if he lived reasonably. Carrying this doctrine to its logical conclusion, certain
social philosophers came to believe that physicians would
not be much in demand in a well-governed society. Indeed,
Plato wrote in his *Republic* that the need for many hospitals
and doctors was the earmark of a bad city. At the most,
physicians were of use only for the treatment of wounds
and during epidemics. "To stand in need of the medical art
through sloth and intemperate diet . . . obliging the skillful sons of Asclepius to invent new names of diseases, such
as dropsies and catarrhs—do you not think this abomina-
ble?" he asks. In Imperial Rome, Tiberius declared in a
similar mood that anyone who consulted a doctor after the
age of thirty was a fool for not having yet learned to regulate his life properly without outside help. In fact, the view
that ministering to the sick is far less important than helping society to maintain health is a recurrent theme of philosophers. "The ancient sages did not treat those who were
already ill," wrote the Yellow Emperor in his *Classic*. "They
instructed those who were not ill. . . . The superior physician helps before the early budding of the disease has already developed." Twenty centuries later Cyrano de Ber-
gerac took a similar view of the practice of medicine in the
utopia that he imagined on the moon. "In every house there
is a Physionome supported by the state who is approximately what would be called with you a doctor, except that
he only treats healthy people."

The concept that disease results from failure to behave
according to natural laws accounts in part for the fact that
illness is more often accompanied by a sense of guilt than
are other misfortunes. Even if he cannot identify his errors,
the patient is likely to experience something akin to shame

arising from a subconscious sense of responsibility for his fate. In Samuel Butler's *Erewhon,* disease was regarded as a sin against society and a man was convicted in court for having developed pulmonary tuberculosis. During the eighteenth and the early nineteenth century, as mentioned in Chapter I, this attitude led Rousseau and his followers to the naive belief that civilized man could recapture the physical well-being and mental virtues of the noble savage by living again according to the ways of nature. Despite its romantic appeal, there is no indication that this literature had any influence in modifying individual behavior and in reconverting sophisticated Europeans into children of nature. But there is no doubt that it fostered an intellectual climate which helped the philosophers of the Enlightenment and the practical sanitarians who followed them to transfer the Hippocratic teachings from the individual to the social level. Out of this attitude arose the social reforms which contributed to the partial solution of the health problems in nineteenth-century Europe.

The shift from the individual to the social point of view in the approach to medical problems was much accelerated by the increase in collective diseases that became apparent in the factories and tenements brought into being by the Industrial Revolution. Several types of motivation contributed to the interest in the crowd diseases that became so prevalent during the nineteenth century. There was probably some realization on the part of the captains of industry that the control of disease among the workers and their families was essential to an adequate supply of efficient labor. More influential, one would hope, was the strong humanitarian feeling which brought into the field of public health the social philosophers and the so-called utopians such as Robert Owen, Saint-Simon, and Fourier. Many of the physicians who turned their attention to the disease problems among the working classes were motivated more by social considerations than by medical interest. Among the first in date was the Italian Bernardino Ramazzini, whose *Treatise on Diseases of Tradesmen* appeared in 1700. In his bulky *Medizinische Polizei,* Johann

Peter Frank proposed at the end of the century that the state was responsible for the people's health. In France, Villerme depicted the hellish conditions prevailing in the cotton mills as an illustration of the harm done by bad industrial practices on the health of labor. J. J. Virey devoted his "Philosophical Hygiene" to problems of health as related to nature, to sociopolitical conditions, and to moral factors.

Rudolf Virchow deserves special mention at this point because of his immense prestige as experimenter, scientist, and writer in several medical and other biological fields. During his student days, Virchow had been influenced by the political philosophy of the German Social Democratic party. At the age of twenty-six in 1847, he was appointed member of a commission organized by the Prussian government to study the epidemic which was then raging in the industrial districts of Upper Silesia. In a minority report Virchow traced the origin of the epidemic to unfavorable meteorological conditions. Heavy rains had ruined the year's crops and this had resulted in famine. Furthermore, the winter following had been extremely severe, forcing the poor people to huddle together in their homes, cold and hungry. It was then that typhus had broken out, first spreading rapidly through the poor population and eventually reaching the wealthier classes. His experience in Silesia led Virchow to start in 1848 the new journal, *Medizinische Reform*. In it he professed, as did his French predecessors inspired by the philosophers of the Enlightenment, that poverty was the breeder of disease and that it was the responsibility of physicians to support social reforms that would reconstruct society according to a pattern favorable to the health of man. In his words, "Epidemics resemble great warnings from which a statesman in the grand style can read that a disturbance has taken place in the development of his people, a disturbance that not even a carefree policy can long overlook." Thus, according to Virchow, the treatment of individual cases is only a small aspect of medicine. More important is the control of crowd diseases which demand social and, if need be, political action. In this light medicine is a social science.

Despite its vigorous intellectual and social basis, the early nineteenth-century health movement in France and Germany was rather ineffective in the way of practical reforms. This failure was due in part to the fact that the goals of the French and German philosophers and physicians were to a large extent political and therefore difficult to reach except by revolutionary action. Furthermore, their doctrines were presented to the public in somewhat abstract terms and not as a concrete message that could be readily understood. In England, by contrast, the leadership was taken by practical men who succeeded in finding a formula that appealed to elementary emotions and was meaningful to everyone.

England had known first and on the largest scale the prosperity but also the terrific destruction of human values that accompanied the first phase of the Industrial Revolution. Laymen as well as physicians could not fail to recognize that disease and physical frailty were most common among the poor classes. In *Conditions of the Working Man in England* Engels spoke of the "pale, lank, narrow chested, hollow-eyed ghosts," riddled with scrofula and rickets, which haunted the streets of Manchester and other manufacturing towns. If ever men lived under conditions completely removed from the state of nature dreamed of by the philosophers of the Enlightenment, it was the English proletariat of the 1830's. To a group of public-minded citizens guided by the physician Southwood Smith and the engineer Edwin Chadwick it appeared that, since disease always accompanied want, dirt, and pollution, health could be restored only by bringing back to the multitudes pure air, pure water, pure food, and pleasant surroundings.

This simple concept was synthesized in the movement "The Health of Towns Association," the prototype of the present-day voluntary health associations throughout the world. Its aim was to "substitute health for disease, cleanliness for filth, order for disorder . . . prevention for palliation . . . enlightened self-interest for ignorant selfishness and bring home to the poorest . . . in purity and abundance, the simple blessings which ignorance and negligence

have long combined to limit or to spoil. *Air, Water, Light.*"
The association undertook a program of education in all
phases of welfare, recommended steps for improving the
attractiveness of habitations and surrounding areas, and
went so far as to make one of its objectives the keeping
open of the lanes about large cities for the enjoyment of
the public. Faith in the healing power of pure air, with
much contempt for the germ theory of disease, was also the
basis of Florence Nightingale's reforms of hospital sanitation
during the Crimean War. "There are no specific diseases,"
she wrote. "There are specific disease conditions."

In Germany the most successful health reformer of the
nineteenth century was the chemist Max von Pettenkofer,
also an opponent of the germ theory. Like his English contemporaries, Pettenkofer regarded hygiene as an all-em-
bracing philosophy of life, implying that not only an abundant supply of clean water and air but also trees and
flowers would contribute to the well-being of men by satisfying their aesthetic longings. He persuaded the Munich
city fathers to have clean water brought in abundance from
the surrounding mountains and to dilute the city sewage
downstream in the Isar. With these steps the great cleaning
up of Munich began and the typhoid mortality fell from
72 per million in 1880 to 14 in 1898. Munich thus became
one of the healthiest of European cities, thanks to the efforts
of this energetic hygienist who was entirely uninfluenced by
the germ theory of disease.

Munich is not the only example of the success of the
Great Sanitary Movement, based in theory and practice on
attempts to reconstitute for a society corrupted by civilization some of the wholesome and aesthetic qualities of ancient days. All contemporary observers report that the general state of health in Europe and North America improved
markedly during the second half of the nineteenth century.
Much of the improvement that followed the removal of
"filth" resulted of course from a decrease in acute infectious
diseases. Barcelona and Alicante, for example, did not experience further yellow fever after the antifilth campaigns
of 1804 and 1827. Everywhere sanitary measures were ac-

companied by a decrease of typhus morbidity and mortality. What gives special interest to these achievements is that they must be credited to the antifilth programs organized by boards of health and other municipal bodies which did not believe in contagion, let alone in the germ theory of disease.

Even tuberculosis, the Great White Plague, showed signs that it could be forced to recede by good and healthy living. At first scrofulous children were sent to the seashore. Then the mountains came into their own. And finally it became obvious that rest and good food in any salubrious and pleasant surroundings were beneficial to many tuberculous patients. It is no accident that the pioneers of the rest cure called *sanitarium* and not *sanatorium* the special institutions devoted to the treatment of tuberculosis. Sanitarium comes from the root *sanitas* and implies healthy living in a salubrious and pleasant environment. It remained the word of choice as long as faith in the healing power of nature prevailed. Sanatorium, from *sanare* (to treat), replaced sanitarium when more active forms of treatment based on the germ theory became the vogue.

Because the decrease in death rates appeared obvious to everyone after 1900, scientific medicine and the germ theory in particular have been given all the credit for the improvement of the general health of the people. The present generation goes still further and now believes that the control of infectious diseases dates from the widespread use of antibacterial drugs. So short are medical memories! In truth the mortality of many other infections had begun to recede in Western Europe and North America long before the introduction of specific methods of therapy, indeed before the demonstration of the germ theory of disease.

Much statistical information is available to document the distant origins of the progress in the control of infection, but two examples will need suffice. The mortality caused by tuberculosis in Europe and North America has been falling continuously and almost at a steady rate ever since the middle of the nineteenth century. From a high point of approximately 500 per 100,000 population in 1845 the mortality had come down to less than 200 at the turn of the century and to 50 in 1945, a tenfold decrease. Yet no drug therapy was available during this period, vaccination was not practiced, and the few therapeutic procedures that were available had but limited value and reached only a very small percentage of the tuberculous population. The decrease in the severity of measles presents an equally startling picture. No technique of vaccination, no drug, no therapeutic procedure is as yet known to deal with this disease. Nevertheless, the accumulated knowledge of old, experienced physicians confirms the statistical information that the disease is much less of a problem now than it was a few decades ago. Clearly the monster of infection had been reduced to a shadow of itself by the time scientific medicine provided rational and specific methods for its control. The conquest of epidemic diseases was in large part the result of the campaign for pure food, pure water, and pure air based not on a scientific doctrine but on philosophical faith. It was through the humanitarian movements dedicated to the eradication of the social evils of the Industrial Revolution, and the attempt to recapture the goodness of life in harmony with the ways of nature, that Western man succeeded in controlling some of the disease problems generated by the undisciplined ruthlessness of industrialization in its early phase.

The Magic Bullets of Medicine

The germ theory, and more generally the doctrine of specific etiology of disease, broke for almost a century the spell of the Hippocratic tradition The core of the new doctrine was that each disease had a well-defined cause and that its control could best be achieved by attacking the causative agent or, if this was not possible, by focusing treatment on the affected part of the body. This was a far cry from the emphasis placed by ancient medicine on the patient as a whole, and on his total environment. The contrast between the two points of view manifested itself in a dramatic manner during the controversies stimulated by Pasteur's communications before the Paris Academy of Medicine. Whereas Pasteur attested that the discovery of specific causes heralded the end of old medicine, his opponents replied that, like firemen fighting a blaze, microbists and chemists were more likely to bring down the patient than to cure the disease. In a less picturesque manner a similar controversy was going on at the same time in German medical circles. Emile von Behring regarded Virchow's belief that misery was the breeder of disease as an antiquated expression of the vague nineteenth-century Naturphilosophie. He asserted that this approach to medical problems could not lead to any effective method of control and opposed to it the concrete knowledge resulting from Robert Koch's discoveries of specific microbial agents of disease.

Von Behring and his like had no doubt that all important infections would eventually be controlled by the use of therapeutic serums and prophylactic vaccines specific for each and every type of microbe. The fashion has changed and for the past decades drugs have occupied the center of the stage in the minds of scientists and medical practitioners, as well as of the lay public and manufacturers of biological products. Whatever the nature of the disease, the most important task—so at least is the well-nigh universal belief—is to discover some magic bullet capable of reaching and destroying the responsible demon within the body of the patient. There are many reasons for believing that the fashion will again change in a not too distant future. As was the case for most serums and vaccines, some disenchantment is bound to follow a critical appraisal of the optimistic promises and glowing reports of early therapeutic and prophylactic success. For the time being, however, the conquest of a disease—microbial or otherwise—is almost always identified with the discovery and use of a drug.

The search for drugs was in the past a purely empirical venture. And, despite lofty attempts at a rational approach to this problem, its greatest achievements are still the re-

sult of chance or at best of trial and error. Indeed, the
procedures presently employed for discovering new drugs
have a strange conceptual similiarity to those used in the
prescientific era. Consider, for example, the history of aspirin—that least celebrated and most useful of all remedies.
Having noted that rheumatic pains were most frequent
among people living in low wet areas, the Rev. Edward
Stone postulated that God in His mercy had certainly
placed in these same areas some antidote for the pains. Inspired by this faith, he discovered in 1763 that an extract
of the willow bark was indeed highly effective in relieving
the pains of rheumatism. Within half a century chemists
had established that the willow extract owed its therapeutic
efficacy to a substance which they called salicylic acid,
from the Latin name of the willow, *salix*. Salicylic acid was
synthesized by Gerland in 1835 and its derivative, acetyl
salicylic acid, by Gerhardt in 1853. The latter substance
was marketed under the name aspirin and for reasons still
unknown proved even more useful than salicylic acid itself.

The faith which guided the Rev. Mr. Stone in his search
for an antidote to rheumatic pain was a manifestation of
the widespread belief that each substance and each effect
has its opposite in nature—a belief which has stimulated
many strange efforts in the never-ending search for drugs.
For example, the mode of thinking which has guided bac-
teriologists in the discovery of antibiotics has a venerable
antiquity, even though it has often led to erroneous conclusions. One story concerns aconite, a plant alkaloid, also
known as love poison, much used in ancient times to dis-
pose of unwanted persons. It is told that Guy di Vigevano,
physician to the court of France, performed in 1335 an experiment from which he believed he had obtained an antidote to the aconite poison. Inspecting closely the aconite
plant, he found its leaves covered with worms and slugs
which were feeding on them. He collected the worms and
slugs, compounded them into a medicine, and fed both
poison and drug to various animals, some of which survived. Convinced by the results of his tests that the medicine was an effective antidote against aconite, di Vigevano

repeated the experiment on himself. He took some of the poison and as soon as he began to feel its effects he ingested the worm-based concoction. According to his own account, he recovered fully, although not without vomiting a number of times.

It might be worth commenting on the fact that the extract of worms and slugs cured the disease caused by aconite at the cost of creating severe nausea in the patient— much as treatment with antibiotics not uncommonly creates a new disease while curing the one for which it was intended. More pertinent here, however, is the method of discovery used by Guy di Vigevano. The faith that a worm feeding on the leaves of the aconite plant should provide a cure for the disease caused by the aconite poison bears a strong resemblance to the theory that a microorganism capable of destroying another microbe on an agar plate should yield a drug effective in the treatment of disease. It is a remarkable fact that this concept, so unsophisticated as to pertain almost to primitive philosophy, has led to some of the greatest practical achievements of modern medicine. But, while great discoveries can come out of primitive concepts, it is true, nevertheless, that science proceeds further when guided by a rational theory. Fortunately, there are a few facts which can be used to provide a rational basis for research on chemotherapy.

In the *Mémoire sur la fermentation appelée lactique*, his very first publication on the germ theory, in 1857, Pasteur pointed out that onion juice did inhibit the growth of the lactic acid ferment but was without effect on the activities of certain other microorganisms. Thus, he clearly recognized that it was possible to inhibit selectively certain types of microbial growths by the use of chemical substances devoid of inhibitory effect on other microbial species. Pasteur never elaborated this point of view and it was left to Paul Ehrlich to develop it into a scientific doctrine.

Ehrlich conceived the thought that effective antimicrobial drugs would be found among substances possessing a selective chemical affinity for some exposed cellular constituents of the parasite to be attacked—the cellular receptors, as he called them. It was this point of view that he summarized in the two famous sentences, "Only such substances can be anchored at any particular part of the organism which fit into the molecule of the recipient combination as a piece of mosaic fits into a certain pattern . . ." and "Antibacterial substances are, so to speak, charmed bullets which strike only those objects for whose destruction they have been produced." More specifically, Ehrlich put forward the suggestion that drugs might act by competing with some of the essential metabolic processes of the parasite and hence by interfering with its nutrition. He proposed this remarkable conceptual scheme as a guide for the development of drugs effective not only against bacteria and protozoa but also against cancer.

Unfortunately, the knowledge concerning the hypothetical "specific cellular receptors" postulated by Ehrlich has not yet progressed very far. Nevertheless, the theory led him to the discovery of several useful drugs, salvarsan being the most famous among them. Even more interesting evidence of the scientific validity of his views is the fact that they have continued to expand and eventually have received a precise formulation in the form of the theory of metabolic antagonism, which is being widely used as a guide to the development of new synthetic drugs.

In practice the drugs discovered either by accident or through a semirational approach must be tested by trial and error in order to establish their practical value. This clumsy and costly way will remain the only one available until a really valid scientific theory of drug action becomes available. But it is a testimony to the power of systematic empirical research that, even in the absence of a good theoretical basis, this wasteful method has yielded such a large variety of drugs useful in the treatment of many types of diseases.

The faith in the magical power of drugs often blunts critical senses and comes at times close to a mass hysteria affecting scientists and laymen alike. The common use of the word "miracle" in referring to the effect of a new drug reveals that men still find it easier to believe in mysterious

forces than to trust in rational processes. Success in all call-
ings is facilitated by the ability to inspire faith and to behave as though part of a priesthood. It is true that faith in
the healing power of ancient gods has somewhat weakened,
but faith itself has lost no ground to reason. Men want miracles as much today as in the past. If they do not join one
of the new cults, they satisfy this need by worshiping at
the altar of modern science. There are always men starved
for hope or greedy for sensation who will testify to the healing power of a spectacular surgical feat or of a new miracle
drug. They provide the testimonies of the new religions for
which scientists with theories unproved or incomplete are
always ready to provide the mystic language. The faith in
the magic power of drugs is not new. In the past, as today,
it contributed to give medicine the authority of priesthood
and to re-create the glamour of ancient mysteries.

Drugs and the Conquest of Disease

It is generally assumed that the discoveries which led to
the use of antibacterial drugs during the past two decades
constitute scientific feats that transcend those of the past
both in theoretical and in practical importance. In reality,
as we have seen, these achievements do not mark the beginning of a new era, but are merely advances along a road
that medicine has been traveling for countless centuries. For
example, the introduction of the cinchona bark for the treatment of fever in the seventeenth century and the discovery
during the nineteenth century that its effective constituent,
quinine, has a suppressive effect on malaria parasites constitute achievements equal in scientific quality and in practical consequences to any of those made during the more
recent era.

The provincial attitude of our time with regard to these
discoveries comes in part from the unequal importance of
various parasites in different parts of the world. Malaria,
other protozoan infections, and worm infestations are the
source of physiological and economic misery in most under-

privileged areas. On the whole, however, the wealthy countries of the Western world suffer relatively little from these afflictions. Except in time of war or when his financial interests are involved, the white man is but mildly interested in diseases with which he has little personal contact. In contrast, his selfishness makes him endow with scientific glamour any discovery that bears on his own well-being. Millions upon millions of human beings in Asia, Africa, and Latin America suffer and die every year from hookworm disease, African sleeping sickness, or malaria. Yet discovery of a drug effective against these maladies has little chance of making headlines, whereas any fact having to do with diseases of importance to the Western world becomes a sensational event. The antibacterial drugs introduced since 1935 owe their unique importance to the fact that they are effective against some of the diseases which were most important for the white man in the Western world a few decades ago.

To be regarded as miracles, furthermore, events must either have occurred in the very remote past or have been published in yesterday's newspaper. Among antibacterial drugs only penicillin and the various mycins developed during or after World War II still rate as miracles, but the sulfonamides which date from 1935 are beginning to be regarded as antiquated drugstore articles. As to the therapeutic achievements of the pre-1930 era, they receive only the lip service paid to the irrelevant truths in some dusty old tome.

It is easy to see how the appearance of the new antibacterial drugs on the medical scene gave rise to the illusion that the age-old problem of infection had finally been solved. A few diseases almost universally fatal could now be cured—subacute endocarditis and certain forms of bacterial meningitis, for example. The course of other infectious processes could be interrupted with incredible rapidity—as in the case of acute streptococcal infections, pneumococcal pneumonia, bacillary dysentery, gonorrhea, syphilis, etc., etc Surgical sepsis became a rarity, thus widening the potentialities of the surgeon's skill. It is obvious that these triumphs of modern chemotherapy have transformed the practice of medicine and are changing the very pattern of disease in the Western world, but there is no reason to believe that they spell the *conquest* of microbial diseases. While it is true that the mortality of many of these afflictions is at an all-time low, the amount of disease that they cause remains very high. Drugs are far more effective in the dramatic acute conditions which are relatively rare than in the countless chronic ailments that account for so much misery in everyday life. Furthermore, as we have seen, the decrease in mortality caused by infection began almost a century ago and has continued ever since at a fairly constant rate irrespective of the use of any specific therapy. The effect of antibacterial drugs is but a ripple on the wave which has been wearing down the mortality caused by infection in our communities.

Most concepts concerning the nature, epidemiology, and control of microbial diseases were formulated during the nineteenth century. This was a time of widespread and killing epidemics, either introduced from the outside, as were cholera and yellow fever, or bred from misery and unsanitary living conditions. Very much the same state of affairs prevails in many parts of the world today, but, on the whole, the great plagues of the past have been brought under some form of control. However, it is only because scientists still think in nineteenth-century terms that these achievements are regarded as having brought about the final conquest of microbial diseases. The appalling number of deaths caused by these diseases was in everybody's mind during the period when the germ theory began to yield its fruits, and it was natural that the lowering of mortality should be the first goal of the medical and social campaigns organized against infection. Now that this goal has been reached, the time has come to realize that mortality rates do not constitute adequate yardsticks for measuring the importance of medical problems. If one were to use as criteria the amount of life spoiled by disease, instead of measuring only that destroyed by death, or the number of days lost from pleasure and work because of so-called minor ailments; or merely the sums paid for drugs, hospitals, and doctors' bills, the toll exacted by microbial pathogens would seem very large indeed. Microbial diseases have not been conquered. Rather, physicians and scientists have resigned themselves to the belief that a relative protection against them can be bought at the cost of a huge ransom.

One might assume that the persistence of microbial diseases is merely a temporary situation, a problem soon to be solved by the discovery of new and more powerful drugs. In reality, there are limitations inherent in drug therapy even under the most favorable conditions. Some of these limitations are technical and cannot be discussed here. Others are more fundamental in character, having their basis in the very philosophy of disease control.

As we have seen and shall again discuss in the following chapter, the characteristics of the total environment—physical and social—determine in a large measure the types of diseases most prevalent in any given community. The belief that disease can be conquered through the use of drugs fails to take into account the difficulties arising from the ecological complexity of human problems. It is an attitude comparable to the naive cowboy philosophy that permeates the wild West thriller. In the crime-ridden frontier town the hero, singlehanded, blasts out the desperadoes who were running rampant through the settlement. The story ends on a happy note because it appears that peace has been restored. But in reality the death of the villains does not solve the fundamental problem, for the rotten social conditions which had opened the town to the desperadoes will soon allow others to come in, unless something is done to correct the primary source of trouble. The hero moves out of town without doing anything to solve this far more complex problem; in fact, he has no weapon to deal with it and is not even aware of its existence.

Similarly, the accounts of miraculous cures rarely make clear that arresting an acute episode does not solve the problem of disease in the social body—nor even in the individual concerned Gonorrhea in human beings has been readily amenable to drug therapy ever since 1935; its microbial agent, the gonococcus, is so vulnerable to penicillin and other drugs that the overt form of the disease can now be arrested in a very short time, and at a very low cost. Yet gonorrhea has not been wiped out in any country or social group. The reason is that its control involves many factors, physiological and social, not amenable to drug treatment. These factors range all the way from the ill-defined conditions which allow the persistence of gonococci without manifestation of disease in the vagina of "successfully" treated women to the economic and psychological aspects of the social environment which favor loose sexual mores and juvenile delinquency.

Other limitations of drug therapy can be illustrated by the attempts to control bovine mastitis, a disease in which the udder of dairy animals can be infected with various types of bacteria. As streptococci used to be the microbes most commonly found in bovine mastitis, it was thought that control of the disease could be readily achieved by treatment with penicillin. It was soon recognized, however, that elimination of streptococci was often followed by appearance of other types of bacteria which took their place in the udder. Mastitis cannot be controlled merely by attacking the bacteria associated with a particular outbreak of the disease. Control can be achieved only by changing the practices of animal husbandry which permit the bacteria to become established and to multiply in the udder. Translated into terms applicable to human diseases, it signifies that drugs cannot be effective in the long run until steps have been taken to correct the physiological and social conditions originally responsible for the disease that is to be treated.

It is a remarkable fact that the greatest strides in health improvement have been achieved in the field of diseases that responded to social and economic reforms after industrialization. The nutritional deficiencies that were so frequent in the nineteenth century have all but disappeared in the Western world, not through the administration of pure vitamins but as a result of over-all better nutrition. The great microbial epidemics were brought under control not by treatment with drugs but largely by sanitation and by the general raising of living standards. In contrast, the cancers, the vascular disorders, the mental diseases, which were not affected by the sanitary movement, have remained great health problems and their solution is not yet in sight. It is legitimate to hope, of course, that vigorous research will yield drugs for the relief of patients suffering from these diseases, but it can be predicted that drug treatment will not provide the real solution to the problem.

The need is to discover and to reform those aspects of the physical and social environment which have brought about an increase in the prevalence of the diseases peculiar to our time. Atmospheric pollution, the exposure to certain chemical agents, the defects in nutritional regimens, the competitive way of life, etc., all have been implicated and probably play some part in disease causation. Many surprises certainly remain in store. The one characteristic of our civilization is the rapidity with which it changes all our ways of life, without too much, if any, concern for the long-term effects of the changes. Man can eventually become adapted to almost anything, but adaptation demands more time than is allowed by the increased tempo at which changes are presently taking place.

The relation of insulin to diabetes illustrates another type of difficulty in the control of disease by the use of drugs. Thanks to insulin, many millions of diabetic persons all over the world can now live long, happy, and useful years. Unfortunately, effective control of the symptoms of diabetes is not synonymous with cure of the diabetic patient, let alone with conquest of the disease. Even when adequately treated with insulin, the diabetic individual is at risk of developing vascular disorders during old age. Still more important from the social point of view is the fact that his children are likely to inherit a tendency to the disease. Thus, the very effectiveness of insulin therapy is bringing about an increase in the prevalence of diabetes in our communities, and a time may soon come when it will prove necessary to weigh the distant consequences of this biological situation. If the tendency to diabetes should become a frequent trait in human beings and if the need for insulin or for an adequate substitute should continue to increase, society may face medical, economic, and ethical problems for which it is not prepared at present. Fortunately there is evidence that treatment with insulin or with drugs having a similar action is not the only possible approach to the control of diabetes. It has been observed that the disease is on the whole less severe in situations where there is a shortage of food, for example, as normally occurs in Asiatic countries or as occurred during the war in the parts of Europe that were under German occupation Clearly the problem is one that demands the most sophisticated medical statesmanship. Its solution transcends treatment of symptoms in the individual patient and might require social reforms reaching even into the field of ethics.

Orthobiosis

There has never been a dearth of opinion as to the rules men should follow to retain health. Every person either has a personal theory on the subject or finds one ready-made in the countless health fads that have existed in all countries at all times. Famous scholars, even medical scientists, have been just as prone as laymen to endorse systems based on half-truths for the proper government of life. Witness the great biologist Elie Metchnikoff, who provides a model in which we can recognize our friends and ourselves.

During the first phase of his scientific life Metchnikoff achieved fame through his fundamental studies on infection and immunity, which won him the Nobel prize. After the age of forty-five, however, peculiar philosophical concepts focused his attention on the problem of old age. He wondered why men fear death and often show anxiety at its approach. Every function calls into play an instinct of satiety. A satisfying meal leaves us without desire for further food; we look forward to rest after sufficient exertion, whether it be work or play. Why, then, do we not experience a desire for death at the end of a normal life? It is, Metchnikoff thought, because human life is usually much shorter than the number of years of which it is potentially capable. Human beings who reach a really ripe age—say 100 years or more—do welcome death without regrets even though they are not sick and do not suffer, just as the normal person welcomes sleep at the end of a full day.

It appeared to Metchnikoff that civilization exerts on modern man some deleterious influence which, although it may not be the cause of overt disease, prevents his life from being long enough to bring into play the instinct of death. For no very good reason, Metchnikoff came to regard the intestinal bacteria as responsible for a slow intoxication causing or accelerating the degenerative diseases of man. He pointed out that parrots and certain other birds which have a short intestinal tract and defecate at frequent intervals usually live much longer than would be expected from their size. The remarkable longevity of Bulgarian peasants was due—according to him—to the fact that they consume large amounts of fermented milk rich in lactic acid bacteria which are antagonistic to the microbial activities responsible for intestinal intoxication. With these thoughts in mind, Metchnikoff reached the conviction that control of intestinal putrefaction by the proper diet would help in preventing degenerative diseases or at least in retarding their course and would thus allow the instinct of death to become manifest. Despite the naive care that he took of his food, Metchnikoff died in 1917 at the age of seventy-one in uremia with extreme manifestations of arteriosclerosis, but he remained convinced to the end that he would have been a healthier man had he started early enough to live according to his theories.

Metchnikoff's fantasies on the relation between intestinal putrefaction, degenerative diseases, and longevity had no basis in fact, either experimental or clinical. His imaginings are now all but forgotten and survive only in the fad for certain fermented dairy products, from the Caucasian yoghurt and kumiss to the sophisticated bacteriologically pure acidophilus milk. Interestingly enough, however, some recent findings suggest that there was some truth in his view that the microorganisms of the intestinal tract may be responsible for a variety of deleterious effects. It has been recently shown, for example, that addition of antimicrobial drugs to the diet of young pigs and chickens increases their rate of growth as well as their efficiency in converting food into meat. Antimicrobial drugs can also protect experimental animals from certain forms of hepatic cirrhosis of dietary origin. Although the exact mechanism of these effects is still uncertain, there is reason to believe that directly or indirectly they involve an inhibition by the drugs of microbial activities which normally result in mild undetected toxemias.

Metchnikoff's obsession with intestinal toxins was tempered by broader wisdom; and he recognized that many aspects of life other than feeding habits had to be reformed if man was to achieve his potential longevity. It was to symbolize all the reforms needed for a long life that he invented the word "orthobiosis"—right living—to encompass all the factors that may affect longevity and well-being. To advocate orthobiosis is of course a counsel of perfection, a restatement of the problem rather than a guide to its solution. Methodologically, furthermore, there is danger in adopting too broad a biological point of view in the study of disease: the danger of substituting meaningless generalities and weak philosophy for the concreteness of exact knowledge. One of the most important contributions of the doctrine of specific etiology was to save medicine from the morass of loose words and vague concepts. But insistence on concrete facts need not deter from acknowledging that, under natural conditions, the etiology of most diseases is multifactorial rather than specific. By using a broadened concept of etiology, encompassing intrinsic and extrinsic determinants of disease, the scientific physician can hope to develop a therapeutic approach that will incorporate the human wisdom and empirical skill of the traditional medical art. "The variable composition of man's body hath made it an Instrument easy to distemper," Francis Bacon wrote. "The Office of Medicine is but to tune this curious Harp of man's body and to reduce it to Harmony."

VI.

SOCIAL PATTERNS OF
HEALTH AND OF DISEASE

The History of Diseases

The ancient Greeks believed that health was the greatest of gifts; St. Hildegarde believed that God did not abide in healthy bodies. Thus, mankind has taken many divergent views of medical problems, but it has never been indifferent to them. Diarists and novelists are prone to discuss phenomena of disease and to explain them in the light of their times and own prejudices. Folklore and learned medical treatises, books of advice to the sick and guides for the preservation of health, and accounts of epidemics and of miraculous cures provide endless material for the history of medicine in all ages. In contrast, medical historians have added surprisingly little to this documentation. They have written a great deal and very learnedly about the lives of physicians and scientists. They have reported in minute details the observations, the experiments, and the theories which have led to progress in understanding disease phenomena and to practical applications in therapeutics and prophylaxis. In brief, they have thrown much light on the development of man's knowledge of disease. But they have written little on the history of the diseases themselves. This neglect gives the impression, and often is certainly based on the belief, that diseases have remained more or less constant throughout human history and that what has changed is only man's knowledge of them

In fact, the available evidence seems at first sight to support this belief. Illness is almost certainly coexistent with life and there is no doubt that ancient man suffered from many, if not all, of the diseases that now plague humanity. For the prehistoric era, bones are of course the most useful surviving witnesses of disease, and their testimony is clear. Rickets, in the past always an accompaniment of deficient insolation, has been identified in neolithic bones of Denmark and Norway. Arthritic diseases were then as today among the most crippling afflictions of man. Thousands and perhaps millions of years ago they occurred in Neanderthal man, in the neolithic caves of France, under the sun of Egypt, and in the mountains of Peru. In fact early man does not occupy a special position in this regard, for arthritis deformans has also been found in dinosaurs, in Eocene and Pleistocene mammals, in a Miocene crocodile and a Pliocene camel. Arthritic deformations have been recognized so often in remains of cave bears that the name "cave gout" has been devised for the disease, the "Hohlengicht" of Virchow. Although malignant tumors of the bone are rather uncommon in prehistoric man and animals, they have been found in Egyptian mummies as far back as 3400 B.C., in human remains of France, North America, and Peru, as well as in fossil horses and cave bears. In contrast, evidence of dental diseases has been found in only a very few fossil animals, but studies on human neolithic skulls in France have revealed an incidence of some three to four per cent of caries. Similar findings made at the prehistoric site of Tepe Hissar in Iran in a population dating from 4000 B C. to 2000 B.C. make clear that, despite common belief, caries afflicted man long before the advent of soft food and candy.

Additional information on disease in ancient man has been derived from X-ray photographs of mummies and from paintings and sculptures. If the interpretation that has been made of these documents is correct, King Siptah of the Nineteenth Egyptian Dynasty and some of his subjects suffered from paralytic poliomyelitis. The first evidence of arteriosclerosis of the aorta has been found in Merneptah, the Pharaoh of the Exodus, who lived about 1200 B C.; hardened arteries also occurred in other Egyptians from the Eighteenth to the Twenty-seventh Dynasty, as well as in

later Greek and Coptic mummies. In fact, the list of diseases
found in mummies reads almost like the catalogue of a
pathological museum and includes silicosis, pneumonia,
pleurisy, kidney stones, sinusitis, gall-stones, cirrhosis of the
liver, mastoiditis, appendicitis, meningitis, smallpox, leprosy,
malaria, tuberculosis, congenital atrophy of the liver. Schis-
tosomiasis, a parasitic disease of backward nations, was
prevalent in Egypt two thousand years ago. The Egyptains
were also plagued by lice and the Peruvians by sand fleas,
as judged from the lesions on the soles of their feet.

The paleopathological records leave no doubt therefore
that most of the known organic and microbial disorders of
man and animals are extremely ancient. But it is also certain that the comparative prevalence and severity of various
diseases have changed greatly in the course of historical
times. Unfortunately, evidence of these changes is never convincing, since the information must be derived from diagnoses at best uncertain and from historical records often in-
accurate and incomplete. Words which appear precise have
had very different meanings at different times. The famous
plague of Athens described by Thucydides was not the
disease known today under this name; it may have been
measles or, more probably, typhus. The instructions to the
lepers in Leviticus certainly referred to a variety of minor
skin ailments in addition to true leprosy caused by the Hansen bacillus. Not all pathological conditions reported as
phthisical or consumptive in old writings can be attributed
to tuberculosis. The word "gout" also has been associated
with a host of diseases—affecting even the stomach and the
heart. William Pitt the Elder had attacks of insanity which
were euphemistically called gout in the brain. Even the
meaning of the word "gonorrhea" has been questioned recently, and the view has been expressed that the disease
referred to under this name by ancient authors was in reality
spermatorrhea. "Gonorrhoea is an unwanted excretion of
semen," Galen wrote, "which you might also call involun-
tary, or to be more precise you might say a persistent excretion of semen without erection of the penis." That Galen
was not referring to an inflammation of the urethra is made

plain by his statement that "Gonorrhoea is an affection of
the organs of seed, not of the pudenda, which are organs
for excretion of seed."

Granted all the difficulties in assessing the significance of
old medical documents, it remains probable that some ten-
tative conclusions are justified concerning changes in the
prevalence of disease in the course of time. There is little
doubt, for example, that the thousands of lazarettos which
existed in Europe during medieval times had been founded
to shelter and segregate lepers infected with the Hansen
bacillus. On the other hand, it is also certain that true
leprosy disappeared almost completely from Europe in the
sixteenth century and that it has never re-established itself
in the countries of Western civilization, although it still constitutes a major disease in other parts of the world.

Another malady which seems to have disappeared or at
least to have changed its manifestations is the English sweating sickness. Between 1485 and 1551 several outbreaks of
this mysterious disorder struck England and to a lesser degree the Continent. Beside the "sweat" even the dreaded
plague paled into insignificance, as can be seen in the diaries
and accounts of the Tudor era. Then the sweat vanished,
for unknown reasons, as swiftly as it had come and it has
not returned to this day—at least not in a recognizable form
—despite the fact that the plague remained rampant until
the middle of the nineteenth century. It is possible that the
sweat was an exotic disease such as dengue, but more probably its microbial agent is still present in our communities
masquerading perhaps as an attenuated influenza or adeno
virus.

In general, changes in incidence of diseases are less sudden than was the case for leprosy and for the sweating sickness. The common picture is that the disease progressively
becomes less prevalent and less destructive but persists in
a milder form, as was the case of syphilis. There is still
much uncertainty concerning the origin of the epidemic
of syphilis that spread through Europe during the sixteenth
century, but there is no doubt concerning its ferocity at that
time. The incidence of the disease was tremendous in all

strata of European society and the lesions that it caused were of extreme severity, as described by Fracastoro and Ulrich von Hutten. From all accounts in literary and medical writings, however, a spontaneous decrease in the virulence took place during the succeeding centuries, although the disease continued to be widespread through all social classes. Observations reported by Bartolomé de Las Cases in his *General History of the Indies*, written in the sixteenth century, indicate that it was then common among the Indians but caused them little trouble. Even more convincing evidence of the spontaneous attenuation of syphilis was obtained some three decades ago. In a village of the Punjab nearly every living person was found to give a positive Wassermann reaction although no stigmata referable to the infection could be detected.

The spontaneous evolution of diseases manifests itself also in the ebbs and flows that occur independently of any conscious intervention of man. Scarlet fever, measles, and mumps serve as convenient examples to illustrate these fluctuations, because of the relative ease with which they can be diagnosed.

In a book printed in 1701 scarlet fever was referred to as "this Name of a disease, for it is scarce anything more." However, the celebrated Irish physician Graves took a different view of the subject a century and a half later. "In the year 1801," he wrote, "scarlet fever committed great ravages in Dublin, and continued its destructive progress during the spring of 1802. It ceased in summer, but returned at intervals during the years 1803–4, when the disease changed its character . . . either so mild as to require little care, or so purely inflammatory as to yield readily to the judicious employment of an antiphlogistic treatment. . . . The experience derived from the present epidemic (1834–35) . . . has proved that, in spite of our boasted improvements, we have not been more successful in 1834–35 than were our predecessors in 1801–2." In 1840 the mortality due to scarlet fever suddenly doubled in England and Wales. From then until 1880 it was the chief cause of death among the infectious maladies of childhood and accounted for four to six per cent of deaths at all ages. The highest mortality from the disease in England was reached apparently during the ten years 1861–70. In 1863 the death rate from scarlet fever was 1,500 per million. At that time, of every million children born in Liverpool 27,000 died of the infection before attaining the age of five.

After this devastating spell the severity of scarlet fever declined and, by 1900, first measles and then diphtheria surpassed it as a cause of death. As is well known, it retained some of its malignancy until the beginning of the present century. As stated by Major Greenwood, it was then *the* fever and required no qualifying adjective any more than the plague does. More recently scarlet fever has again become a relatively mild disease. Indeed, it is probable that much of the credit for its control, which is commonly given to drugs and improved methods of treatment (as was done at the beginning of the nineteenth century), should in reality go to the unknown factors which brought about the "spontaneous" decrease in its virulence.

A similar story can be told of the fluctuation in the severity of measles. William Heberden asserted in 1785 that measles "are usually attended with very little danger; it is not often that a physician is employed in this distemper." But the position changed sharply about 1800. In 1804 measles caused as many deaths, chiefly among adults, as did smallpox and actually surpassed the latter in 1808. The illness appears to have resembled what Sydenham in 1674 called "anomalous" or "malignant" measles and it is probable that its mortality was even greater than indicated by statistics since it must have caused many fatal chest affections that remained undiagnosed. Measles remained the main cause of child deaths until around 1840; then it began to ebb and scarlet fever displaced it for over forty years as the chief infectious disease. But toward the end of the century measles again resumed some of its importance and almost every year until 1915 the deaths from it outnumbered those from smallpox, scarlet fever, and diphtheria combined. From then on its mortality fell and has continued to fall ever since.

We have repeatedly used the expression "spontaneous" to qualify those changes in the prevalence and severity of diseases that have occurred without the conscious intervention of man. In reality, of course, these changes are the expression of natural forces; for example, the hereditary factors concerned in evolutionary adaptation that have been considered in earlier chapters. Another aspect of the problem is the influence that socioeconomic factors have exerted on the prevalence and severity of disease. As we shall presently see, each type of civilization has had diseases peculiar to it and at each period the various social groups in any community also have differed in this regard.

Hunger and Surfeit

In their fundamental outline the nutritional requirements of man are the same all over the world and have probably not changed in the course of history. But men's nutritional needs cannot be defined merely in terms of calories, proteins, fats, sugars, and vitamins. Ancestral habits and cultural influences limit greatly the type of foods that are acceptable and often impose the use of others which are actually deleterious. In one form or another malnutrition is almost as common in the midst of plenty as undernutrition is in times of shortages. "They are as sick that surfeit with too much," says Nerissa in The Merchant of Venice, "as they that starve with nothing."

Hunger and sex have always been among the basic drives of men and animals. To judge from poetry, novels, and plastic arts, however, hunger is no longer a preoccupation of Western man. The discoveries of psychoanalysis seem to give scientific sanction to the view that sex and its sublimation, love, now account for most of human actions and sufferings. In reality it is because the middle and upper classes of nineteenth-century Europe had all the food they wanted that hunger could be ignored by their psychoanalysts and that sex came to occupy such a unique place in their preoccupations. Hunger recovered its place as the dominant drive during World War II in the concentration camps as well as in the population of the occupied countries—usually bringing to the surface bestial aspects of human nature even in the most civilized men.

In many parts of the world starvation is still a constant threat which affects emotional behavior and imposes characteristic patterns on the social structure. As the food supply of primitive people usually gets low before the start of the growing season, individuals are liable to become listless at that time and the life of the group falls to a low ebb. The sense of excitement that comes with spring originates not only from the physical well-being caused by warmer weather and brighter light but even more from a sense of expectancy for the new crops. The orgiac celebrations that marked the cult of Dionysus and Demeter in Greece were in essence biological outbursts associated with the exhilarating oncoming of the new growth. Even today, in most of us, there emerges on the first warmer days some feeling that the genial processes of nature are again at work, as if some spirit of life were really circulating and generating new strength in our bodies.

Because of their dramatic character, periods of famine loom large in history. However, famines do not constitute the most interesting aspect, nor the most important, of the effects of nutrition on human affairs. As already mentioned, the diet of many primitive peoples is limited in quantity and extremely monotonous during the late part of the dormant season, whereas it may be varied and abundant after harvest-time. The Andaman Islander will live for a time upon pork, next upon fish, then have to resort to honey and wild fruit. The African may have a quantity of fresh vegetables available at a season, then pass to a diet of mushrooms and fruit at another, then be limited to porridge and dried peas and beans. Few are the natives who limit themselves to a vegetarian diet out of choice. They do it out of necessity, as shown by the enormous amounts of meat that they are likely to consume at a sitting when game becomes available.

These alternating periods of nutritional scarcity and

plenty cannot help having profound physiological consequences, but these have been little studied except for their obvious effects. Anthropologists who have lived among primitive people report that sudden changes in diet often result in severe gastrointestinal disturbances which the natives usually accept as a matter of course. These obvious effects, however, are probably less important than others of a more subtle nature associated with the physiological adaptations required by profound changes in nutritional regimens. This situation is rarely encountered now in the Western world, since a varied supply of food is generally available throughout the year, but what appears on the surface as a great advantage may also have some deleterious effects. Man, like wild animals, evolved under conditions in which the quantity and nature of food varied from one season to another and these cyclic alterations of scarcity and plenty may be reflected in certain physiological rhythms that are now ignored.

Nutritional deficiencies commonly occur as indirect results of social and technological changes. Primitive cultures manage to derive fairly adequate nourishment from even the most desolate areas if the ecological conditions under which they live are stable. Thus, the Otomi Indians have found sufficient sustenance to survive for many generations in the Mesquital valley of Mexico, a semidesert with many months of drought yearly. They exhibit little, if any, clinical evidence of malnourishment even though they consume few of the foods usually considered essential for a balanced diet. Their consumption of meat, dairy produce, fruit, and conventional vegetables is extremely low. Along with their tortillas, they eat purslane, cactus fruit, pigweed, and sow thistle for greenstuffs, and drink pulque. Yet their diet when tested a decade ago was found to supply a better nutritional combination than that eaten by certain city dwellers in the United States examined at the same time. It is probable that every component of the Otomi diet provides some essential nutrient. The alcoholic beverage pulque, which is used unfiltered, contains all the factors of the agave juice from which it is made, as well as those generated by the microorganisms involved in the fermentation process.

The fine adjustment achieved by primitive people is, however, extremely precarious and is readily upset by any change in their economy. We have already mentioned the ill effects that resulted from the replacement of millet by corn in the diet of Zulus and other Bantu-speaking Africans. It is often claimed that the incidence of dental caries increases rapidly as primitive people change from their native to European diets. Likewise, the substitution of polished for unpolished rice in the Orient has had unfortunate consequences, the nutritional mechanism of which is now fairly well understood. In rice the vitamins and fats are present in highest concentrations in the superficial layer of the kernel, just below the silver-fleece. As these substances encourage the multiplication of bacteria and insects, silver-fleece rice is rapidly infested with mites and other pests and usually becomes rancid within a few weeks. Whole rice, therefore, is a practical source of food only for rural populations, and is at its best for only a short time after harvesting. Polished rice can be more readily stored and for this reason has displaced the unpolished form, but it is likely to produce nutritional deficiencies when it becomes the main dietary item.

Inadequacy in protein intake is the most important cause of malnutrition in the world today. Cirrhosis of the liver, one of its consequences, is extremely common even among children in tropical countries, and extreme susceptibility to many types of infection and infestation is another of its consequences. In the Western world the problem of dietary protein is now solved to a large extent by the availability of foods of animal origin. But these are expensive to produce and the type of farming which they require is beyond the economic possibilities of underdeveloped countries, especially in times of rapid population increase as at present. In theory it is possible to prepare nutritionally adequate mixtures of plant proteins, supplemented, if need be, with fish, but this has not yet been achieved practically in a form acceptable to human taste.

Not so long ago food shortages were also common among the poor classes in Europe, and in addition lack of nutritional knowledge certainly resulted in nutritional deficiency even among the rich. Rickets, for example, was so prevalent in England during the seventeenth century as to pass for a normal state. Charles I was affected, like other children, and his physician merely reported that the "joynts of his knees, hips and ankles being great and loose are not yet closed and knit together as it happeneth to many in their tender years which afterwards when yeares hath confirmed them prove very stronge and able persons." The swaddling of infants certainly originated in the prevalence of rickets and also as a measure to prevent premature weight bearing and the development of deformities. The increase in size of the European people during the past few decades suggests that the nutritional regimen in the past rarely permitted maximum growth of men and women even among the favored classes. As judged from the armor and costumes on display in museums, medieval knights and the aristocracy that followed them were much smaller than the average American soldier of World War II. Likewise, the eighteenth-century belle was but a diminutive version of today's debutante.

Several comparative studies made of young people whose families emigrated to the United States and of their kin who stayed in Europe or in Oriental countries reveal an astonishing difference in size between the two groups. As any traveler knows, American children look like young giants in comparison with their Oriental counterparts. The average size of children is increasing also in most parts of Europe. A recent report from Glasgow shows that the boys are taller by almost 4 inches at age 13 than they were 40 years ago; the corresponding gains for girls has been 3¼ inches. The boys weigh 14½ pounds and the girls 16½ pounds heavier than the children of 40 years earlier. Similar findings have been made in other parts of the Western world. In London choirmasters have been made aware of the increased growth rates in children by the fact that, according to a recent editorial in *The Lancet*, they are finding it increasingly difficult to get choirboys with a soprano voice.

This acceleration of growth during the early years of life is probably the result of many independent causes. More sensible clothing and control of infectious diseases have probably played their part, but most important certainly is the change in nutritional regimen. Under normal conditions the growth of the modern child in the Western world need never be interrupted by seasonal nutritional deficiencies, as was the case a century ago and as still occurs among nonindustrialized people. At the present time in America seasonal changes in the rate of growth are observed almost exclusively among Indian children living on the reservations. In all other population groups the availability of cow's milk as well as of a great variety of food throughout the year permits continued rapid growth by providing readily digested, nutritionally complete proteins.

There is no doubt, therefore, that man has developed the technological and nutritional know-how for rapid gain in height and in weight. But it has not yet been proved that the biggest baby, vitamin- and nitrogen-stuffed, is necessarily the best baby. Nor is it certain that enjoying three square meals a day with a constant supply of all nutritional factors at all seasons of the year constitutes an unmixed blessing. There exist in the body of man, as of all animals, biological mechanisms for the storage of food developed for meeting the irregularities and cyclical changes in nature. It may still turn out that a nutritional way of living permitting continuous growth at a maximum rate may have unfortunate distant results. Fasting fads may have some justification after all by providing an opportunity for the operation of certain emergency mechanisms built by nature into the human body.

It is only during recent years that overnutrition has been recognized as a form of malnutrition. In laboratory experiments rats fed an unlimited diet were found to die sooner than animals prevented from gaining weight by a diet severely restricted in quantity but well balanced in composition. Likewise, insurance statisticians have repeatedly emphasized that in man the obese have a short expectancy of life. Indeed, obesity is now publicized as the most common nutritional disease of Western society. This was apparently true also in Imperial Rome, as during all periods of great material prosperity. "In the old days," wrote Lucretius in the fifth book of *De Rerum Natura,* "lack of food gave languishing limbs to Lethe; contrariwise today surfeit of things stifles us." Thus, history repeats itself. Like the prosperous Romans of two thousand years ago, countless men of the Western world today are digging their own graves through overeating.

The Diseases of Pestilence and of Sanitation

War and Famine have long been known to ride with Pestilence. And likewise other aspects of the political and social history affect the type, prevalence, and severity of microbial diseases.

The most dramatic examples of pestilence have been provided by the introduction of pathogenic agents into populations which had not had recent exposure to them. Thus, it is widely recognized that increase in international trade brought plague to the Roman world of the Justinian era and again to Europe during the Renaissance. This simple statement fails to convey the complexity of the social factors which have governed in the course of history the dramatic interplay between man, the plague bacillus, rats and other rodents, and the various types of fleas.

The destructive effects of plague, pneumonic as well as bubonic, on any affected community were so sudden and so completely paralyzing—killing at times a large percentage of the population within a few months—that all its manifestations have long been recorded in great detail. As a result, much knowledge has accumulated concerning the effects of the physical environment and of social factors on the genesis and course of plague epidemics. The causative bacillus was discovered in 1894, and the characteristics that account for its ability to cause disease are on the whole well understood. It is known also that the plague bacilli commonly infect rodents, particularly rats, and can be transmitted to man by fleas that abandon the body of the dying animal Detailed studies have been made of the types of rats and fleas involved in the transmission of the bacilli and of the conditions under which the disease is most likely to spread through human communities.

There are now several procedures to deal with the epidemics—by controlling the rodents, by eliminating fleas, by protecting exposed populations with vaccines, by treating the sick patient with antimicrobial drugs. All this precise knowledge acquired through laboratory experimentation is essential but not sufficient to understand the history of plague. For the relationships between man, rats, and fleas are influenced by countless external factors ranging from the vagaries of the weather to the whims in vestimentary fashions. The weather is affected by the sunspot cycle, the crops are affected by the weather, and rodent population, and therefore plague, by crop abundance. The world is one and forms part of a cosmic unity.

The main reservoirs of the plague bacilli in nature are wild rodents which are infected but sufficiently resistant not to suffer from their infection under normal circumstances. They carry the plague bacilli throughout their life, just as so many healthy men and women are infected with tubercle bacilli or with viruses without showing signs of disease. Among the naturally infected animals are the tarabazan (Manchurian marmot), which has long been hunted for its fur. The professional Manchurian hunters carefully avoid any tarabazan which appears to be sick and in fact a religious taboo specifically instructs them in this regard. This taboo is probably related to the fact that the plague bacilli become active in sick tarabazans and therefore can more readily be transmitted to man. Around 1900 there occurred a change in women's fashion in Europe which increased the demand of the fur trade for the pelt of the tarabazan. Attracted by the high prices of the fur, inexperienced Chinese took to tarabazan hunting. Being ignorant of the ancient taboo, they did not hesitate to catch sick animals, which proved the easiest prey. Several of the hunters caught plague from the tarabazans and transmitted it to the population in the inns of Manchuria. Thus began the great epidemic of pneumonic plague in Manchuria.

Changes in the rat population brought about in part by human influence played an important part in the ebbs and flows of plague epidemics. Plague became rare in Europe approximately at the time when the brown wander rat (*Rattus norvegicus*) began to oust the domestic black rat (*Rattus rattus*) from buildings in the majority of European cities. The wander rat apparently swarmed across the Volga in 1727 and thence westward over Russia. By the end of the century it had spread by land and sea and displaced the weaker, less prolific, and more home-loving black rat as the common species associated with man. Related to this change in rat population is the fact that the flea *N. fasciatus* is most common on the wander rat, whereas the flea *X. cheopis* prefers to infest the black rat. As *X. cheopis* is a far more effective transmitter of plague of man than is the other flea, there is reason to believe that the change of rat and rat-flea species during the eighteenth century was of paramount importance in the disappearance of plague from Europe. Since, on the other hand, the initial swarming of the wander rat was probably the consequence of some obscure phenomenon, perhaps climatic in origin, it becomes extremely difficult indeed to identify the "causes" of the epidemic waves of plague.

Still other factors of social origin are relevant to the problem. As the wooden and loam houses of the Middle Ages, with their roofs of straw or rushes and their many warm nooks, began to disappear, they were replaced progressively by stone and concrete structures without hiding places, which were much less congenial to the black rat. During recent decades, however, a general extension of the area of the black rat has been observed on the European continent and in the British Isles. The increased transports from southern regions during the two world wars may have been responsible for this change, as the black rat is much more liable to be transported than the brown rat. Immigration, however, is only a part of the problem, because recent changes in building methods and living habits are once more providing conditions favorable to the black rat population in Western Europe. The lining of roofs with boarding affords them living accommodations with mild temperature assured by central heating. The radio and TV aerials, as well as telephone wires, provide them with many new means of access which are not used by the wander brown rat. Furthermore, the installation of kitchens on the attic floor makes available sources of food in spots eminently suited to the black rat. By reason of its nature and habit the brown rat does not benefit by these changes and in fact modern ratproofing of buildings is an important obstacle to its maintenance. Increase in the black rat population need not, of course, mean a danger of new plague epidemics, since so many other factors are involved in the transmission of the bacillus from rat to man. It means, however, that control of the disease may require more careful surveillance than at the time when the brown rat had displaced the black rat in our communities.

During the sixteenth, seventeenth, and eighteenth centuries, as we have seen, the European navigators, conquerors, and settlers left as poisoned calling cards wherever they set foot a host of microbial agents which were part and parcel of their own racial history. The introduction of tuberculosis, scarlet fever, smallpox, measles, etc., proved to be catastrophic events in the life of the Amerinds and the Polynesians when these peoples first came into contact with Europeans. Similarly, poliomyelitis caused havoc a few years ago among the Greenland Eskimos who acquired the virus, new to them, from well-meaning European or American visitors.

The white man has been so gregarious and has traveled so much in the past two millennia that he has had occasion to come into contact with a large percentage of his potential microbial enemies and has developed through selection and immunization some form of racial resistance to many of them. But, just like primitive people, he falls an easy prey to infection when the circumstances are right. After all, spectacular epidemics strike only now and then. In Europe leprosy was prevalent in the fourteenth century, plague in the fifteenth, syphilis in the sixteenth, smallpox in the seventeenth and eighteenth centuries, scarlet fever, measles, and tuberculosis in the nineteenth century. No one was prepared for the fury of the influenza pandemic when it struck after World War I Who can say what is in store for the future and how effective the modern methods of prophylaxis and treatment, the vaccines and drugs, would have proved in the face of these killing epidemics to which Western man had not developed any natural resistance at the time they reached him.

Clearly, the mere introduction of the microbial agent is rarely sufficient to establish an epidemic state, and every epidemic could probably be shown to have been conditioned by some aspect of the social climate. It is not through accident that the signs of tuberculosis are described frequently and at length in the ancient literature of India, Greece, and Rome and of other complex civilizations with crowded cities, whereas the disease is not referred to in the Bible or in other lore of pastoral peoples. Tuberculosis illustrates particularly well the relation between ways of living and prevalence of disease because its fluctuations are clearly determined by economic upheavals and follow a course that reflects social history

The epidemic of tuberculosis that spread through the Western world during the nineteenth century was the outcome of the social tragedies that followed in the wake of the Industrial Revolution The need for labor had brought about a huge and sudden shift of population from rural to industrial areas, an extraordinary migration of people which gave to Verhaeren's *Les Villes Tentaculaires* and to Goldsmith's *The Deserted Village* their dramatic atmosphere. In the mushrooming cities, the migrants found dreadful working and living conditions. Long hours of exhausting toil were exacted of them in the suffocating atmosphere of coal mines, in the dark factories, and in the damp offices. Child labor was ruthlessly exploited, particularly in the textile mills. Malnutrition prevailed in the filthy, crowded tenements, and the bleakness of life was relieved only by gin and vice.

Within a few decades millions of individuals raised on farms and in small towns were thus uprooted and exposed suddenly to the debilitating effect of inhumane employment. Living in slums and fed on bread, porridge, and potatoes with very rarely some cheese and even more rarely a small bit of bacon, they constituted the proletariat bred by the early phase of the Industrial Revolution all over the world. In the United States it was the fate of the Irish immigrants driven by the potato famine out of their homeland and into the crowded mill towns of the Atlantic coast. The hardships caused by industrialization were not limited to poor lodging, inadequate food, and physical exertion. Another of its evils was to rob millions of human beings of the values and emotional satisfactions which had made their lives bearable in the past.

Most of the new recruits to industrial labor had known poverty in their former rural surroundings, but their life had been relatively free from stresses. More important, they had achieved some sort of physiological and psychological adaptation to their humble social status. They had enjoyed the sun, birds and flowers of their villages; they had learned to adorn the dullness and drudgery of existence with bright ribbons and jolly tunes, and with the pageantry of their church. When they moved into industrial areas in search of prosperity, adventure, and comfort they found instead the anonymous gloom of the industrial cities, with little chance to escape from squalor and despair. And while under stress, before having adjusted themselves to their new ordeals, they came into contact with city dwellers among whom tuberculosis had long been prevalent. Intense crowding in workshops and in unsanitary living quarters provided all that was required for the rapid spread of infection, while physiological misery favored the development of destructive disease. It is this constellation of circumstances which brought about the explosive spread of tuberculosis among the laboring classes, and from this huge focus the infection spread through society by means of countless unavoidable contacts.

The association of tuberculosis with rapid industrialization can once more be observed in several parts of the world at the present time. Large-scale migrations from rural areas to urban districts, low standards of living, disruption of ancestral habits—all these earmarks of the Industrial Revolution are found in much of Latin America and of Asia. Concomitantly, tuberculosis is manifesting itself in these regions with the huge death rates that prevailed in Europe and the United States in the 1830's.

Along with tuberculosis, many other infectious diseases spread throughout the Western world during the Industrial Revolution. But just as the numbers of deaths caused by tuberculosis began to decline in Western Europe and North America around 1850, so did the mortality caused by other pulmonary and intestinal infections. Clearly, the decline in mortality was due in part to biological and social forces independent of the introduction of specific therapeutic measures. England and America, which had been the first to become industrialized, were also the first to recover from the epidemic, whereas countries in Latin America and Asia, which are still in the initial phase of the Industrial Revolution, are now experiencing the full fury of the epidemics. In any given country the death rates of infections are liable to soar to a high peak shortly after the shift from a rural to an industrial type of economy; then the epidemics lose their acute character and their mortality falls as prosperity becomes more widespread.

It is to be hoped that social upheavals will never again occasion the physiological misery which allowed the killing epidemics of the nineteenth century, but other social factors may increase the importance of infectious processes. Chronic bronchitis, for example, is becoming one of the most frequent causes of disability in some parts of the world. As a disease it does not have the blood-curdling character of the great pestilences of the past and for this reason tends to be ignored by sociologists and historians. Nevertheless, its prevalence in certain areas of Northeastern Europe reveals unmistakably that the disease corresponds to a well-defined epidemic climate. In England bronchitis ranks in importance as a cause of death with arteriosclerotic heart disease and is the cause of some fifteen per cent of premature retirement from work on grounds of ill-health. In 1951,

37,000 people died of bronchitis, and the disease accounted for the certified loss of 26.6 million days among the insured population in England. Like coronary heart disease and cancer of the lung, bronchitis affects chiefly the adult and older men and accounts, with the other two diseases, for most of the ominous middle-age excess in contemporary male mortality.

Bronchitis exhibits a striking geographical gradient. It is rather uncommon in Norway and Denmark, slightly more common in Sweden, France, and Switzerland, and much more frequent in Western Germany, the Netherlands, and Belgium, with a peak in the British Isles in general and in England and Wales in particular. The rate of occurrence seems to be related to industrial and domestic smoke pollution, being highest in the black areas where the smog of 1952 was the most intense. In addition to being exposed to bad weather and to industrial dust and fumes, the labor population in these areas lives in defective and overcrowded homes with high risks of multiple infections. In both sexes the incidence of bronchitis is five times higher in the least favored social class than in the most favored. Thus, it is clear that the prevalence of chronic bronchitis depends upon a combination of climatic circumstances, air pollution, and social factors, which increase susceptibility to various types of infection of the respiratory tract.

Intestinal disorders of various sorts were the dominant filth diseases of past centuries, on the whole, this kind of filth has now been cleared from our streets and gutters, but industrial civilization has brought in a new kind of dirt which pollutes the air and thereby lifts infection from the intestinal to the respiratory level. No drug, however potent in antimicrobial activity, can control the infections associated with bronchitis and sinusitis and no vaccine can protect against them, just as it is almost certain that no drug and no vaccine could have controlled the intestinal diseases of filth in the nineteenth century. It is filth that must be dealt with, or rather a social attitude that must be changed. Each civilization has its own kind of pestilence and can control it only by reforming itself.

In 1803 Malthus wrote in his *Essay on Population,* "I feel not the slightest doubt that, if the introduction of the cowpox should extirpate the smallpox, we shall find . . . increased mortality of some other disease." Events have proved him correct, for measles and then scarlet fever took the place of smallpox as cause of death. Some sixty years after Malthus, William Farr wrote in his annual letter to the Registrar General in London, "The infectious diseases replace each other, and when one is rooted out it is apt to be replaced by others which ravage the human race indifferently *whenever the conditions of healthy life are wanting.* [Italics mine. R.D.] They have this property in common with weeds and other forms of life, as one species recedes another advances." Farr's pessimism found its justification in the ravages of typhoid and influenza during the next half century.

Just as happened in the past, social changes in the future are liable to bring in their train microbial diseases of an unsuspected nature. Indeed, there are a few signs that the sanitary policies themselves, the very techniques which contributed so much to the control of infection in the late nineteenth century, are presenting mankind with new problems of infection. As long as the viruses of poliomyelitis were distributed so widely as to infect every child soon after birth, at a time when the immunity conferred by the mother's blood was still at a high level, paralytic poliomyelitis was extremely rare. With increased sanitation, however, first contact with the virus is likely to be delayed and as a consequence infection is contracted later when maternal immunity has disappeared, with the result that the paralytic and fatal forms of the disease are becoming increasingly common in adults. Thus, the increase in incidence and severity of poliomyelitis, which is noted chiefly in the most sanitary nations, seems to be the unexpected outcome of effective plumbing and social cleanliness. German measles, likewise, used to be a universal disease of childhood without serious consequences for health. In our communities, however, many children are now protected from contact with the virus and thus escape infection during childhood.

As a result of this fact many people of the new generation
reach adulthood without having had a chance to develop
immunity and thus may contract infection later in life. If
a young woman contracts infection during the early stages
of pregnancy, the disease is likely to cause congenital mal-
formities in her child. Thus, German measles, which was
but a trivial accident when contracted during childhood,
can now become a major problem at a more critical stage
of life, on account of improved sanitation.

Modern Horsemen of the Apocalypse

War, Famine, and Pestilence still commonly ride in advance of Death in most of the world today. But these time-
honored allies of the pale rider are now less feared than they
used to be. Men may not be happier, nor even fundamentally healthier, than their ancestors but in the Western world
at least life expectancy is longer. The human mind has been
freed of its obsession with death and disease caused by
violence, nutritional deficiencies, or infectious fevers, and it
can now return with more confidence to some of its ancient
dreams of eternal youth and long life.

Death, however, is now acquiring new allies that are taking the place of Famine and Pestilence. Horsemen of destruction that were rarely seen in the past are increasingly
threatening the life and soul of modern man. Vascular
diseases ruin his heart or brain; cancers run riot; mental
diseases break his contact with the world of reason. As was
the case for the great epidemics, two kinds of medical
philosophy are guiding the approach to the control of these
modern endemics. One is the search for drugs capable of
reaching the site of the disease within the body of the
patient. The other is the attempt to identify those aspects
of modern life thought to be responsible for the disease
problems peculiar to our times.

The search for magic bullets against cancer, vascular
disease, and mental disease is made especially difficult by
the fact that so little is known of the target that they must

reach in the body. This problem epitomizes the dilemma encountered by medical scientists concerned with the search for cures. They try to apply rational methods to the discovery of drugs, but realize that this is a counsel of perfection rarely compatible with practical exigencies. Life is short, the art is long, the problems pressing. Even the most empirical approach is justified in medical research if it offers any hope of yielding results of practical usefulness. In the present state of ignorance the only possible method in most cases is the empirical hit-or-miss attack dignified in scientific circles by the expression "screening program."

While the search for the magic bullets continues, other studies are revealing that the environment in which the individual lives and his manner of living are of great importance in determining his susceptibility to the diseases of modern times. The cancerous ulcers of chimney sweeps probably constituted the first convincing demonstration that the incidence of tumors can be increased by certain chemical substances, in this case by certain components of coal tar. Many other substances have since been found capable of eliciting cancer, and one of the most disturbing aspects of this problem is that many of the carcinogenic effects reveal themselves slowly, often requiring years to become manifest. For example, workers who use benzine in the rubber industry often develop leukemia late in life, the solvent persisting long unnoticed in their bone marrow and ganglia. Awareness of these delayed effects is causing much alarm, of course, to those responsible for the safety of foodstuffs, as few of the dyes used in the food industry can be considered entirely safe from this point of view. Furthermore, many other hazardous chemicals find their way into food as thickeners, sweeteners, flavors, preservatives, etc. Unfortunately, the problem cannot be solved by restricting food additives to products of "natural" origin, for many constituents of plants and many other natural products have carcinogenic activity. In fact it has been suggested recently that the high incidence of stomach cancer in certain areas of Holland and England might be related to the composition

of the water originating from the geological strata under-
lying those areas.

Radiations are another factor of the environment involved in the production of tumors. Thus, skin cancers are
unusually frequent in fair-skinned people doing outdoor
work in sunny countries, probably due to the carcinogenic
effect of ultraviolet light in the range of 3,200 A; maxillary
sarcoma used to be common among women engaged in
painting the dials of phosphorescent watches with radium.
Of special interest at the present time is the high incidence
of leukemias among persons exposed to excessive doses of
certain radiations, whether they be physicians in their professional activities, patients receiving radiotherapy, or persons exposed to the effects of nuclear bombing. As in the
case of carcinogenic substances, the effects of radiations are
often delayed; witness the leukemias observed in the physicians who pioneered in the use of X rays or beginning to
appear now among atomic physicists.

Many strange aspects of the distribution of cancers have
recently come to light: the low rate for cancer of the breast
in Japan; the high rates for cancer of the stomach and
esophagus, with low rates for cancer of the lung, in Finland
and Iceland; the effect of marriage customs on the ratio of
breast cancer to cancer of the uterine cervix; the differences
in occurrence of primary liver cancer in Africa, Indonesia,
and Europe. In most cases it is not yet possible to trace
the local occurrence of certain cancers to a particular determinant cause, but this is not surprising because so many
different factors can play a role in cancer causation. In man,
cigarette smoking and air pollution have been implicated in
lung cancers; in experimental animals certain viruses can
certainly pave the way for many different types of tumors.

Ill-defined as they are, all these observations suffice to
suggest that the environment exerts a profound effect on
the relative prevalence of the different types of cancer. In
view of the fact that so many various agents can function
as determinants of cancers—from the radiation of the sun
to the constituents and products of the earth—it is futile to
try to escape from the disease by returning to the ways of

nature. But it is worth while to define more precisely the factors of the environment which contribute to cancer causation, since this knowledge bids fair to facilitate its control through prevention.

For a long time cardiovascular diseases were thought to be the inescapable consequence of aging, the natural manifestation of senescence with its decay of the body structures, more or less rapid but inevitable. In reality, however, arteriosclerosis is not a necessary accompaniment of aging, and blood vessels can remain "young" in structure and in function to a very old age. In 1635 there lived in Shropshire, England, one Thomas Parr who was reputed to be 152 years old. Old Parr was living quietly on his native heath when his fame reached the Constable of England, who ordered him to London. There he was presented to the King, well wined and dined—and thereupon promptly died. An autopsy was performed by none other than the great William Harvey, who found all the organs normal and healthy. No calcification was detected, no anatomical cause of death, and old Parr was declared to have died of a surfeit. He was buried in Westminster Abbey.

There are several other well-authenticated examples of centenarians whose arteries showed no trace of sclerosis. It seems to be a fact, furthermore, that arteriosclerosis does not strike at random, but occurs most commonly in certain parts of the world and certain social groups. Diets rich in meat and certain fats, absence of physical effort, and the pressure in everyday life imposed by highly competitive mores are among the factors which have been implicated as contributory to vascular diseases. Much has been made of the fact that the Neapolitans and other South Italians, with their starchy diet and far niente philosophy, suffer less from coronary diseases than do the Bolognese and other North Italians who eat more fat and exhibit a more aggressive behavior. Observations made during World War II seem to have provided further evidence for the view that abundant and rich diet bears a relation to cardiovascular accidents. In Norway and other occupied countries the deaths from coronary disease and thrombotic accidents became much less frequent when the Germans moved in and the calories and fats moved out, whereas these diseases resumed their prewar level shortly after alimentation became normal again after the war—the 3,600-calorie diet replacing the 1,800-calorie diet. But intriguing as these relationships are and useful in suggesting working hypotheses, they do not as yet constitute convincing evidence of the determinism of vascular disease. A rich table and easy living differ from austerity and hardships in many factors other than those concerned in blood levels of purines, cholesterol, or other lipids.

In addition to the killing diseases, like cancer and vascular disorders, which have monopolized public and scientific attention during recent years, there are many others which illustrate the effect that social habits exert on the state of health. As judged from the dentition in mummies, dental caries was rare in Egypt during the predynastic period and became more frequent with developing civilization, particularly among the wealthy classes. And subsequent periods of Egyptain history provide further evidence that every form of dental disease was more prevalent among the wealthy than among the poor people who had to be satisfied with a coarse, uncooked diet. Similar conclusions have been derived from studies on the children of African tribes. In a comparative investigation the caries rate was found to be twenty-eight per cent among the Luo children, who are town dwellers and eat manufactured foods, whereas it was only eleven per cent among the Banyaruanda children, whose parents are poor and unable to buy European foods other than tea and sugar. Similarly, the aborigines of Australia, who live on kangaroo meat, which is tough and rubbery, have little caries and remarkably healthy paradontal tissues, capable of transmitting a biting stress far greater than those of more civilized man.

Gout has long been used to illustrate the deleterious effects of overindulgence in rich food. In reality little is known concerning the pathogenesis of this disease, despite the fact that the nineteenth-century caricaturists seem to have had well-established convictions on the subject. In countless cartoons they depicted the grimaces of the portly bourgeois or nobleman—especially English—paying for his rich fare of roast and port through excruciating pains in his gouty toe. It has been said in support of the caricaturists that gout became much less common in Great Britain after the war had imposed on all classes the discipline of austerity diets.

Men are naturally most impressed by diseases which have obvious manifestations, yet some of their worst enemies creep on them unobtrusively. For example, any significant addition to background radioactivity increases the mutation rate of all earth's creatures from plant to man, with results that will not be felt to the full for several generations. And, likewise, the continuously mounting pollution of the air and the inescapable contact with drugs and chemicals that are becoming part of everyday life carry threats which are less obvious than cancer or heart disease but at least as important.

Occupational diseases have long been recognized under a number of picturesque names such as chimney sweep's cancer, mule spinner's cancer, miner's phthisis, wool sorter's disease, etc., but it is only during recent decades that industrial processes have grown in variety and magnitude to an extent that affects the life of practically everyone.

The smoke and mist that evoke romantic moods in rural environments are replaced in industrialized areas by the dreaded smogs which kill many within a few days, cause painful irritation to the mucous membranes of all, and may be the source of much unforeseen disease that will become manifest only in the future. In the course of one single episode of smog that hung over London for four and one-half days during late December 1952, almost five thousand persons died—as well as the prize cattle in their stalls at the Smithfield Show.[1] This acute mortality was followed by a

[1] Although the prize cattle at the Smithfield Show were badly affected by the smog, ordinary cattle kept for slaughter were not. The former were kept in beautifully clean stalls, whereas the latter stood in the familiar barnyard atmosphere with a high concentration of ammonia. It has been suggested that this ammonia neutralized both the sulfur dioxide gas and the sulfuric acid mist which contributed much to the irritating character of the smog.

secondary and more prolonged plateau of deaths, some eight thousand above normal up to the end of February 1953. In addition to the irritating substances of which everyone is aware, smogs contain other constituents that may well produce deleterious effects many years later. Smogs are the symbol of the new factors of the environment that man has created and that will constitute threats to himself and his race until he has learned to eliminate them altogether or has succeeded in reaching with them some sort of biological and social adaptation. In the modern industrial world smogs have become "the pestilence that walketh in darkness."

There is perhaps some symbolism, as yet incompletely deciphered, in the visions that St. John the Divine had on the island of Patmos. The first horsemen of the Apocalypse that he saw were Famine and Pestilence. Then another even more terrifying visitation was sent by the angered Deity. After the fifth angel had sounded his trumpet he opened the bottomless pit and "there arose a smoke out of the pit, as the smoke of a great furnace; and the sun and the air were darkened by reason of the smoke of the pit." And out of the bottomless pit came the scorpions that did not kill men but tormented them for five months before final destruction came "by the fire, and by the smoke, and by the brimstone." The time of fulfillment of the Apocalypse may not be far off.

From Madness to Boredom

In the United States during 1956, close to one million persons were hospitalized for mental disease, more than ten million were thought to be in need of psychiatric treatment, and there was reason to believe that a good percentage of the total population would spend at least part of their lives in a mental institution. In 1955, 16,760 persons were known to have committed suicide and not a few of those who had died a violent death had directly or indirectly been victims of abnormal or antisocial behavior. On the other hand, it was considered a great medical advance that tranquilizer drugs had become available to all. Three out of ten prescriptions were for these drugs in 1956, and more than a billion tablets of meprobamate alone were sold in a year!

Psychiatric illnesses are not a problem special to the United States, but really constitute a burden of ever-increasing weight in most of the countries of Western civilization. Like all quantitative statements pertaining to human affairs, however, enumeration of disturbed persons or of psychiatric beds gives but a distorted impression of the change in incidence of mental diseases in modern times. It is true that the number of hospitalized persons has been increasing for many years. But it is also a fact that the village fool who used to be an accepted member of any rural setting, the semisenile oldster who was expected to spend his last years rocking on the porch of the family homestead, and even the timid soul who escaped competition by retiring into a sheltered home atmosphere are likely now to become inmates of mental institutions because they cannot find a safe place in the crowded high-pressure environment of modern life. Thus, the problem of our time may be less an actual increase in the numbers of mental defectives than a decrease in the tolerance of society for them. The sickness of the individual is not readily differentiated from the sickness of society.

Among mental diseases some can be traced to fairly well-defined organic damage, to general paralysis or alcohol psychosis, for example. Others result from individual stresses related to particular life situations. Still others, probably the largest number, correspond to group psychoses associated with certain cultural patterns. The dancing mania and tarantism of the Middle Ages, the madness of the New Guinea natives, and the ghost dances of the Plains and Ute Indians are among the many forms that have been well described. But whatever their origin, all forms of psychosis have a large social component and it is this aspect of the problem that we shall now consider.

Although mass psychoses have been observed all over the world, it is usually assumed that mental diseases are less common among primitive and semicivilized peoples than in highly evolved, complex societies. In reality there is little useful information on this subject. It is always difficult to dissociate ailments with an organic basis from those of psychogenic origin, even when the observer is dealing with people of his own culture, and the difficulty is still greater in the case of primitive people who rarely differentiate between the two classes of disease. The fact that medicine men and magicians so often achieve remarkable therapeutic successes through incantations and suggestions shows that much of disease in primitive societies has a large mental component.

On the other hand, it is likely that the incidence of psychiatric disorders tends to increase whenever a people begins to intermingle with a more complex civilization. It appears, for example, that mental diseases among the Kalmucks and the Kirghizes, who had in the past remained almost completely isolated, increased markedly after 1850 when these tribes first came into intimate contact with Western civilization. A similar state of affairs obtains presently among many tribes of North American Indians One of its expressions is the alcoholism so prevalent among them when they pass from the protective atmosphere of their own culture into the no man's land that they must cross before achieving integration with the white people. The spread of the Peyote cult is probably another aspect of the same phenomenon. The consumption of the hallucinating drug and all the rites associated with it are attempts to replace by a new religion the social traditions lost as a result of disturbance of the ancestral Indian culture.

Psychotic disorders resulting from the process of acculturation are not peculiar to primitive people coming into contact with the white man's world. They occur wherever and whenever social changes are too rapid to allow for gradual, successful adaptation. If psychiatric illnesses are truly increasing in the Western world, the reason is not to be found in the complex and competitive character of our society but rather in the accelerated rate at which old habits and conventions disappear and new ones appear. Even the marginal man can generally achieve some form of equilibrium with his environment if the social order is

stable, but he is likely to break down when the extent and
rate of change exceed his adaptive potentialities. For this
reason mental diseases are likely to become more apparent
in areas undergoing rapid cultural transitions, and this
means a large part of the world at the present time. Man
in undeveloped countries is subjected to the strain of moving
straight from a camel's back to a motorcar or to an airplane
without the intermediate experience of a stagecoach. In the
most industrialized countries, also, electronics and automation are bound to cause maladjustment and social stress by
revolutionizing from one year to the other the ways of life
and the techniques of production, as well as the amount
of leisure and the manners of entertainment. It probably
took Western man long periods of relative stability before
he could enjoy the peace of the Sabbath and the sociable
evenings at the end of the day's work. The four-day week
may be the cause of as many stresses as was the exploitation of labor a few generations ago.

Automation and dial watching eliminate the hardships of
physical effort, but monotonous environments and mechanical operations have their own deleterious effects on the
human brain. Recent studies have revealed that as a result
of prolonged exposure to a monotonous situation "the individual's thinking is impaired; he shows childish emotional
responses; his visual perception becomes disturbed; he
suffers from hallucinations; his brain-wave pattern changes."
In brief, the efficiency of industrial production is creating
a pathology of boredom.

Man, furthermore, is a social animal, and intimate participation in the creative activities of his group constitutes a
much-needed sense of fulfillment that he often had the
opportunity to enjoy in the past. To a large extent this creative satisfaction is now denied him as a result of technological changes. Modern man has acquired the economic power
to own things, but these are anonymously produced. He
may even have some abstract knowledge of their nature and
operations, but ownership and knowledge are not substitutes
for direct experience and do not satisfy the ancient need for
being a real participant in the act of creation. Failure to

satisfy this need breeds at times despair, often boredom, or at least the listlessness so prevalent in modern societies.

The history of man, both racial and social, is a long saga of difficulties overcome, of emergencies that had to be met in order to avoid destruction. Dangers, real or imaginary, and fear of the unknown used to be part of everyday life, but the unexpected also contributed an atmosphere of adventure and expectancy—the type of exhilaration that helps man to free himself from bondage to matter and to reach for the stars. Accumulation of earthly goods does not make up for this exhilaration, without which the zest for life is readily lost. The indifferent and the outwardly satisfied are less likely to retain happiness and mental sanity than are those who sacrifice well-being and comfort for the sake of ideals or illusions.

The highly organized social structures of modern times try to provide the individual with security from the womb to the grave, but this security is often bought at the cost of boredom. The exhilaration of expectancy constitutes in large part the salt of life. Boredom is not easy to define or to recognize, and its onset is insidious. It often masquerades in the passive forms of entertainment, in the dreary hours of aimless driving, in anonymous holidays which have lost their meaning because they are no longer holy, as well as in the attitude of the person who "couldn't care less" about the events of the world around him. Its manifestations go from the various forms of escapism, such as addiction to drugs or alcohol, to suicide, which relieves the victim of the need to care about anything.

According to statistics recently published, the highest rates of drug addiction, suicide, and death from violence occur in countries which enjoy material wealth and have the most effective social legislation and greatest political stability. The United States, Switzerland, Denmark, and Australia led the list for the period 1948–51. With regard to both suicide and homicide (as well, of course, as to road accidents), the lowest rates occurred in countries where nature, economic circumstances, and inadequate social legislation seem to make life harder and more uncertain Among

these countries are Catholic Eire and Spain, Protestant
Ulster and Scotland, and the Jewish state of Israel. It has
been reported that, even under the worst and apparently
most hopeless conditions, the rate of suicide was extremely
low among the internees in German and Japanese concentration camps. It seems as if the will to live—the eagerness
to overcome the varied and inescapable problems of existence—is often weakened by physical comfort and wherever
the cushion of the welfare state has been made too soft and
has provided shelters against most of the vicissitudes of life.
The modern state is predicated on the assumption that
happiness is best reached through freedom from want and
from struggle, whereas evolution-wise life implies strife and
adventure. Modern man is apparently finding it hard to
escape from his biological past.

Medicine and Society

It is now the responsibility of the state to control the
affairs of man in a manner conducive to better health for
the majority. In Europe this paternalistic attitude was codi-
fied in the series of volumes published between 1779 and
1817 by Johann Peter Frank under the appropriate title *A
Complete System of Medical Police*, that covered every
phase of man's life, from municipal planning, housing, sanitation, water supply, and sewage disposal to child hygiene
and school planning. The modern era has seen the fulfillment of Frank's dreams. But while it is true that the general
state of health of the people has improved in the Western
world during the past 150 years, it is not too clear how
this has been brought about. In a lecture delivered in 1855
the English sanitarian Southwood Smith made the brash
statement that "Epidemics are now under our control. . . .
We have banished the most formidable." (Italics mine.
R.D.) To support these claims he could point to the fact
that plague, cholera, remittent fever, malaria, and typhus
had all but disappeared from London following the campaign against filth. A century later one would hear every-

where similar lectures, but boasting now that the conquest
of infectious diseases had finally been brought about by the
use of antimicrobial drugs and promising the early conquest
of cancer through the discovery of other drugs. In reality,
as we have seen, the health of nations is affected by almost
any change in their social pattern—from the drainage of
swamps to the shortening of working hours, from the introduction of cotton undergarments to the refinement of food
products and to the most sophisticated medical practices.

Contrary to common belief, even the much-discussed increase in population has occurred independently of the control of disease. As far as can be judged the world population
was approximately 450 million at the mid-seventeenth
century. It has now reached 2.5 billion, a fivefold rise. But
this increase began 300 years ago, even during the period of
the great epidemics, and it has taken place in many countries which have not enjoyed the benefits of modern medicine.

Life is like a large body of water moved by deep currents and by superficial breezes. We have gained some understanding of the winds and can adjust our sails to them.
But the really powerful forces which determine population
trends are deep currents of which we know little, the fundamental physical and biological laws of the world, the habits
and beliefs of mankind with their roots deep in the past.
It is intellectual deceit to be dogmatic in these matters on
the basis of scientific knowledge, because information is so
incomplete. And it is always dangerous to bring about radical and sudden social changes, because the complexity of
the interrelationships in the living world inevitably makes
for unforeseen consequences, often with disastrous results.
The use of knowledge must be tempered by humility and
common sense, and for this reason medical utopias must
be taken with a great deal of salt. In this light it is important to remember that the control of microbial diseases in
the Western world occurred progressively and was sufficiently slow to permit orderly adjustments. In contrast, the
present efforts to eliminate infection rapidly in the underdeveloped countries by radical public health measures is

almost bound to bring about biological disturbances and to give rise to new population problems before there has been time for achieving compensatory changes in the rest of the environment.

Of course, these considerations cannot influence the behavior of the physician toward each individual patient. But they introduce new types of responsibilities in the problems of medical statesmanship involving whole populations. They illustrate how difficult it is to define and delineate the role of medicine in the community. In the words of a wise physician, it is part of the doctor's function to make it possible for his patients to go on doing the pleasant things that are bad for them—smoking too much, eating too much, drinking too much—without killing themselves any sooner than is necessary. But it is also the doctor's role, as claimed by Rudolf Virchow, to recognize that social disease is the manifestation of a process affecting the community as a whole.

Clearly, health cannot be dissociated from any of the factors that influence human welfare and happiness, and yet it is not the function of medicine to become identified with political action, were it only for the reason that medical training does not necessarily impart to physicians the wisdom and skill required to deal with sociopolitical problems. What Napoleon said of coquettes and of men of wisdom applies to physicians. "It pays to see them, to converse with them, but it is usually unwise to select a wife or a cabinet minister from among them."

Physicians, unfortunately, cannot ignore the economic connotations of the problems of health and disease, for economic exigencies may soon become the limiting factors in the application of medical knowledge to the welfare of mankind. It has been calculated that, in 1949, almost $11 billion were spent in the United States on medical and institutional care of disease; this corresponded to more than five per cent of the national income for that year, to which must be added some two per cent lost by disability to work. A special report to the President also revealed that in 1952 a million American families had spent fifty per cent of their total familial income on medical care, and 8 million were in debt on that account. These enormous figures are alarming not so much by reason of their size but because they represent a trend. As is well known, the very advances in medical science are constantly increasing the cost of medical care, a consequence of greater availability of various therapeutic procedures. On the other hand, there is a limit to the percentage share of its income that a society, however prosperous, can devote to the control of disease. It is a disturbing fact that Western civilization, which claims to have achieved the highest standard of health in history, finds itself compelled to spend ever-increasing sums for the control of disease. Some of our ways of life may not be so favorable to the maintenance of health as our endless boasts would lead one to believe, and it is probable also that much of the effort spent on disease control is misdirected because of faulty formulation of medical policies. Indeed, the cost of medical care constitutes a problem that Virchow would have regarded as requiring "medical statesmanship on the grand social scale."

Above and beyond economic considerations there are aspects of Western culture which are incompatible with the incessant claims that we have come close to the millennium of good health. It is true that we have cleared our dwellings of the filth and vermin that sheltered and nurtured the killing microbes of disease; but we remain as much as ever easy prey to ill-defined microbial ailments which do not take life but just ruin it and which we can neither diagnose nor cure. We have eliminated some of the greatest and most obvious contaminants of food and drink; but we poison our atmosphere, and endanger future generations as well, with the gases of chemical processes, the smoke of factories, the pulverized rubber and exhaust of motorcars, and man-made radiations. We have eliminated from modern society some of the crudest forms of economic injustice; but we operate human relationships on a basis of aggressive competition and of endless striving for success. We have lavishly produced and made available to men a wealth of comforts and earthly goods; but we have denied to most of them the possibility of choosing and of participating creatively in the joy of production. For too many life in the modern world is a passive experience or a lonely struggle, the wounds of which are reflected not only in damage to the blood vessels of our brains and our hearts but also in the very loss of hope.

For many centuries the mass of people derived emotional satisfaction and mental equilibrium from their intimate association with the Church and with its celebrations that followed the rhythms of the year. Now, in the words of D. H. Lawrence, we have "nothing but politics and bank holidays to satisfy the eternal human need of living in ritual adjustment to the cosmos in its revolutions. . . . The human race is like a great uprooted tree, with its roots in the air. We must plant ourselves again in the universe." But in practice everything in the modern world tends to isolate us from the universe, and it is considered a triumph of medical science to have made the process more complete and more rapid by the use of tranquilizers!

A cartoon published recently in *The New Yorker* shows a well-clad, well-fed, obviously prosperous and decent man in an anonymous crowd. In a listless mood he stands in front of a vending machine looking at its display of Miltown, Phenobarbital, Doriden, and Benzedrine. "Get through the day, five cents," the vending machine advertises to the man who is puzzled as to whether his greater need is to be stimulated in order to function more effectively or to be quieted down in order to forget and relax. Clearly it is a form of disease for a society to have become dependent on drugs to be able to "get through the day." The vending machine and its addicts to tranquilizers or stimulants symbolizes a social attitude of gloom and even tragedy. It makes one long for the not-so-distant past when refuge from the threats and ordeals of everyday life could be sought and sometimes was found among wine, women, and song.

Just as the great epidemics of the nineteenth century were precipitated by environmental factors which favored the activities of pathogenic microorganisms, so many of the diseases characteristic of our times have their origin in some faulty factor of the modern environment. It is not to be doubted that precise knowledge of the physicochemical mechanisms of these diseases will eventually provide means for their alleviation, but maintenance of joyous health is a higher goal than discovery of a cure. The significance of the sanitary movement of the 1830's for the history of mankind resides in the fact that it was the first conscious and organized effort not for the treatment of disease but for the creation of a healthier, happier world. Its leaders approached the problems of health with much practical skill, but it must never be forgotten that a philosophical and humanitarian doctrine was the inspiration of their pragmatic genius. A similar ideal might again inspire a new pioneering venture to attack the health problems of the present day.

It is not impossible that in the future, as in the past, effective steps in the prevention of disease will be motivated by an emotional revolt against some of the inadequacies of the modern world and will result from the search for a formula of life more akin to the natural propensities of man. This attitude need not mean a retreat from science —far from it. The crusade for pure air, pure water, pure food was at best a naive and often ineffective approach to the problems of health of the nineteenth century, but it paved the way for the scientific analysis of the factors responsible for the epidemic climate of the Industrial Revolution. Similarly, scientific medicine will certainly define the factors in the physical environment and the types of behavior which constitute threats to health in modern society. But to fulfill its potentialities it may once more need the help of bold amateurs willing to use empirical methods based on philosophical, humanitarian, and aesthetic beliefs. Medical statesmanship cannot thrive only on scientific knowledge, because exact science cannot encompass all the human factors involved in health and in disease. Knowledge and power may arise from dreams as well as from facts and logic. Utopias are often but the memory of Arcadias.

VII

EFFECTS OF DISEASE ON POPULATIONS AND ON CIVILIZATION

Food, Disease, and Population Trends

Wherever conditions are suitable, living things multiply to the fullest possible extent allowed by the availability of food. The number of rats in a city increases with the abundance of garbage in the streets, just as vegetation becomes more abundant when fertilizer is added to the soil. On the other hand, food shortage calls into play mechanisms which tend to limit the number of offspring. In mice kept in their natural habitat under controlled conditions the numbers of matings and the rate of fecundation are little affected by the amount of food supplied, but resorption of the fetus occurs frequently when food supply is restricted and thus keeps down the total population. Many other regulatory mechanisms in nature limit animal populations to levels compatible with the proper development of the individuals that reach maturity.

In the case of man the chain that binds reproduction to abundance of food is often weakened by cultural influences. Nevertheless, it is clear that some of the large spurts in population size that have occurred during historical times have been coincidental with new agricultural practices or trade policies resulting in increased food supply. The introduction of the bean into the Rio Grande Valley brought about a marked increase in the numbers of Pueblo Indians during the sixteenth century. The Navajo tribe increased rapidly in numbers after it began to raise sheep and goats instead of basing its subsistence merely on the products of hunting. Corn, the sweet potato, and peanuts were introduced into China between A.D. 1550 and 1600, providing readily cultivated, large-yield crops which supplemented the traditional grain diet. From approximately 60 million in 1578 the population increased to 100 million in 1660, 140 million in 1741, 300 million in 1850, and approximately 600 million at the present time. So far as is known no other significant change occurred in the Chinese way of life; certainly no new sanitary or medical practice was instituted. Yet the population increased at an accelerated rate despite recurrent bouts of pestilence and of warfare. Similarly, the domestication of the white potato introduced from the Andes coincided with a population spurt in Europe.

In England and Wales the population rose from 10 to 34 million during the nineteenth century—not counting the millions who emigrated. In this case the increase in food supply had its origin in the overseas trade stimulated by the Industrial Revolution. Benjamin Franklin had given much thought to the population problem. In his essay *Observations Concerning the Increase of Mankind and the Peopling of Countries,* he expressed the belief that more Englishmen would be west of the Atlantic Ocean than east of it within 100 years. Neither emigration nor immigration, he felt, was of any significance in affecting the size of population, whereas "If you have room and subsistence enough . . . you may of one make ten nations . . . or rather increase a nation ten fold in numbers and strength."

Extreme hunger will naturally check population growth temporarily. Loss of sexual desire was one of the early effects of malnutrition in concentration camps; the famine in occupied Netherlands at the end of the war was followed nine months later by a marked decline in births. Under ordinary conditions, however, food shortages do not seem to constitute effective checks of population size. The birth rate has remained extremely high in most of Eastern Asia despite the fact that chronic malnutrition and periodic famines have prevailed there for many centuries. Clearly, the restrictive mechanisms that regulate population trends in wild animals fail to operate as effectively in man because of cultural interference. Ancestor worship in China, religious ethics in Catholic countries, the need for family labor in rural communities, and simply fondness for children are all factors which tend to increase the population often beyond what appears to be the biologically optimum level. Whatever their state of destitution, most human communities make every effort to permit the survival of children, even when it is known that these will suffer from malnutrition and disease throughout their lives. One of the most successful creations of the UN has been UNICEF, an agency organized to provide food and medical care to the underprivileged children of the world. "Save the children" is now the most universally accepted motto of mankind. But it is also recognized by all that the increase in natality and the fall in infant mortality will soon bring to an acute stage the dangers of overpopulation in the underdeveloped parts of the world.

In his *Essay on the Principle of Population as It Affects the Future Improvement of Society*, published in 1798, Malthus propounded that, since any population unchecked in its growth increases geometrically while its food supply could only increase at an arithmetic rate under usual conditions, it was unavoidable that a large part of the world would be condemned to live under marginal conditions. While the abundance now prevailing in Western countries appears incompatible with Malthus' law, the present food-population problem in the rest of the world gives a burning actuality to his gloomy views. To save people from death by measures of public health is proving relatively easy, but no solution is in sight for the many problems created by their survival.

Experts point out that all the possibilities of food production have not yet been exploited, and that agricultural and industrial technology will eventually make it possible to feed some 10 to 50 billion people—four to twenty times the present world population. But fulfillment of this goal would require drastic changes in the pattern of life, including the elimination from the earth of practically all wild animals that compete with man for space and for plant products. This process, in fact, would not be a radically new departure, for it has been going on for countless generations. Over much of the planet, animal life is allowed to survive unmolested only if it contributes to the needs and pleasures of man, and to the extent that it does not interfere with his interests and comfort.

There are, fortunately, many sources of food other than those presently in use. Valuable proteins can be prepared from grasses for human consumption. The oceans have hardly been tapped. Practical techniques for fishpond culture and for the growth of algae under factory conditions could be developed rapidly if needed. Synthetic chemistry may eventually render obsolete the conventional methods of food production. Even now chemically produced foods constitute important items of diet. Synthetic vitamins are displacing the vitamins of natural origin; several thousand tons of fats were produced from coal tar for human consumption during the Second World War; and recently some of the essential amino acids have been produced industrially on a large scale and at very low cost, thus promising to relieve man from his dependence on animal products for adequate protein nutrition. The days of synthetic foodstuffs in standardized pills may not be far off.

Hunger, so long one of the main drives of human activity, may thus be eliminated as a biological experience through technological advances in food production. But, while man can develop a technology capable of producing all the calories, proteins, fats, and vitamins that he needs, the freedom from nutritional want may bring in its train nutritional boredom. While the requirements formulated by nutritionists can certainly be met through scientific and technological research, it will be harder to satisfy man's longing for the tasty morsels that he used to relish. Indeed, the prospect of a diet made up of synthetic foodstuffs appears at first distressing and even unacceptable to the gourmet, because the idea of food is identified with certain conventional tastes, shapes, consistency, and color. Fortunately, man pos-

sesses much adaptability and ingenuity with regard to his
culinary habits. He has learned to mask the taste of rotten
fish with mayonnaise and other sauces. He has acquired a
liking for the countless odoriferous kinds of cheese derived
from milk by weird fermentation and putrefaction processes. He will certainly manage to devise acceptable recipes
for the dull produce of synthetic chemistry After all, little
culinary skill will be required to impart to the synthetic
foodstuffs of tomorrow some of the qualities of taste and
structure that no longer exist in the bland and spiritless
breads of today.

It is frequently stated that wars and especially disease
constitute the most effective checks to unrestricted population growth. There is no doubt, indeed, that certain diseases can have this effect by shortening the average span
of life and by decreasing the birth rate. The enormous infant mortality that prevails wherever malnutrition is widespread and sanitation defective is probably the most important factor in this regard. During recent years in Ceylon
the generalized use of the insecticide DDT brought about
a reduction not only in malaria but also in infant diarrhea.
Simultaneously, the over-all death rate dropped almost one-
half and the birth rate rose Thus, one single public health
measure resulted directly or indirectly in a rapid increase
of the Ceylon population. It is far from certain, however,
that the effect of public health on population growth will
be lasting. While it is obvious that malnutrition and infection take a large toll in most communities, it is also true
that various compensatory mechanisms can neutralize their
effects. As a rule, birth rates are the highest precisely in
the areas and groups where infant mortality also is the highest. Even the loss of life caused by wars and epidemics has
in general but very transient effects on population trends.
The tremendous death tolls exacted by plague during the
Justinian era and the Renaissance, or more recently by influenza and the two world wars, were in each case more
than made up for by actual increases in world population
within a few decades. Wherever and whenever a vacuum is

created, human life rushes in to fill all the haunts that it is capable of occupying.

On the other hand, the fact that the world population is increasing presently at an alarming rate does not necessarily mean that this trend will continue indefinitely. The stimulating effect of technological advances, sanitary practices, and improved medical care might well be followed by a phase of falling birth rates, as has been observed on several occasions in European countries. When food shortages and killing diseases are no longer acute problems, the distaste for effort, the desire for comfort, the craving for social prestige, as well as other cultural forces, become limiting factors in population growth. Food may one day be produced in amounts sufficient to feed 50 billion people, but past experience shows that men who are adequately fed develop preoccupations and tastes which tend to restrict the number of their progeny.

Thus, population trends are influenced by many opposite forces. On the one hand, biological exigencies and cultural urges make it appear unwise to let the world population grow beyond a level compatible with a life comfortable, safe, and easy for all. On the other hand, men continue to want children despite the threat of starvation, disease, war, and disaster. Mankind behaves as though governed by the subconscious faith that life is always a worth-while adventure, even when the actual experience spells privations and tears.

Effects of Disease and Nutrition on Military and Political History

Just as the social environment conditions the state of health at a given time in a given community, so in turn problems arising from health and disease are reflected in the activities and the very moods of each particular civilization. Societies, like individuals, may yield passively to their diseases or react constructively against them, but they cannot escape their influence.

Among the effects of disease on the course of history those having to do with war have been most completely recorded and discussed. The subject was made popular by Zinsser's witty best-seller, *Rats, Lice and History,* in which he propounded the theory that the activities of disease germs have been more important than the plans of generals and the leadership of statesmen in determining the outcome of military conflicts. More than two thousand years ago disease made literary history with Thucydides' account of the "plague" of Athens in 430 B.C. during the Peloponnesian War. This epidemic, which killed one third of the city's inhabitants, including Pericles, was not plague, despite the name under which it is known. Thucydides' description has suggested diagnoses ranging from yellow fever to scarlet fever, smallpox, typhus, measles, typhoid fever, dysentery, etc. In contrast, it was almost certainly true plague which struck the Philistines when they captured the Ark of the Covenant from the Israelites.

Smallpox, as much as Roman virtue, contributed to the defeat of Carthage, and it was probably smallpox also which ravaged Rome during the second century A.D. and killed Marcus Aurelius. The fate that befell the Amerinds when they first came into contact with the European invaders is among the best-documented illustrations of the effects that epidemics in general and smallpox in particular have exerted on the course of wars. Shortly after the arrival of Cortez and his Conquistadores, smallpox spread like wildfire through the Indian population. It seems to have been first transmitted to the Indians in 1520 by a Negro in Cortez' band who was in the active infectious stage of the disease. As smallpox was then prevalent in Europe, the Spaniards had probably developed immunity to it through early exposure, whereas the Indians, who had had no racial experience with it, proved very susceptible. By killing at least half the Indians and demoralizing them at a critical time, the epidemic certainly played a part as important as Spanish arms and valor in bringing about the conquest of the South American continent.

History repeated itself in North America. Smallpox,

measles, scarlet fever, tuberculosis, other infectious diseases, as well as alcoholism, decimated the Indians and contributed to the breakdown of both their physical stamina and their morale. Captain Bartholomew Gosnold, who explored the New England coast in 1602 and named Martha's Vineyard, found "the people of a perfect constitution of body, active, strong, healthful, and very witty." Yet within a few years of his visit the Indians from Narragansett to Penobscot had been reduced from 9,000 to a few hundred, probably by smallpox. When the Puritans reached this territory in 1620 they found the bones of the dead unburied. Similarly, the Massachusetts tribe is supposed to have been reduced by infection from 30,000 to 300 in a short time.

In contrast to the Spanish missionaries, who worried over the fact that conversion to Christianity so often proved a death warrant to the natives, and that sickness and death were too frequently the result of their endeavors, the English in North America during the seventeenth century openly rejoiced when the Lord sent his "Avenging Angels to destroy the heathen." In fact the Europeans soon became aware of the fact that smallpox was one of their most effective weapons against the Indians and they did not hesitate to spread the infection intentionally by means of contaminated blankets, always on the pretext that it helped to destroy the enemies of the faith. God is always on the side of the strong battalions, even when they are made up of microbes.

Disease has continued to interfere with the master plans of strategists during modern times. The terrific losses that Napoleon's army suffered as a result of typhus, dysentery, and leptospirosis contracted during the march through Poland and Russia contributed to the disasters of the 1812 campaign as much as did Russian resistance and the hardships of the winter. Similarly, typhus in Serbia, dysentery in Gallipoli, trench fever and influenza on the western front, played their role in the military campaigns of World War I. During World War II, likewise, dysentery and typhoid paralyzed at a critical time the Italian army in Libya, and infectious hepatitis played havoc with the German army and particularly with Rommel's Afrika Korps.

Malaria deserves special mention here because it has so long and so profoundly affected the history of all peoples. It is blamed for the death of Alexander the Great by the waters of Babylon; it is believed to have attacked the Gauls as they besieged Rome and to have upset the Italian expeditions of Lothaire and Frederick I Barbarossa during the Middle Ages. In the seventeenth century it was present among the English troops operating in the Low Countries. In the nineteenth century it played a large role in America during the Civil War, in Europe during the Crimean War, and in the European colonial wars. During World War I all the eastern armies were affected, but it was especially during World War II with reference to the Pacific theater that malaria played an important role in the plans of statesmen and military commanders. By seizing the Dutch Indies early in the war and thus gaining control of the supplies of quinine, the Japanese thought they would paralyze Allied operations in the malarious regions of the South Pacific and Asia. Indeed, this strategic move might have met with success had not the synthetic drug atabrine become available in unlimited quantities to provide an effective antimalarial substitute for quinine. At one time in Burma the casualties from malaria were some ten times those caused by Japanese weapons of war, making the functions of the Director of Medical Services more important for the conduct of the campaign than those of the General Staff. It is estimated that about half a million American soldiers were at some time hospitalized with malaria during that war. Still more recently, malaria was common among the combatants in Korea and Indochina.

In addition to epidemic diseases, other less obvious medical problems have had decisive effects on the issue of battles. Scurvy played its part in the eventual defeat of the Crusaders, just as it spelled the failure of many land and sea explorations. Whereas the Northern armies during the American Civil War received abundant meat and dairy products, the diet of the Confederates was often low in protein and consisted to a large extent of corn products and molasses. On many occasions the Confederate armies failed to exploit and pursue to complete victory military advantages gained early in the day, as if the Southern soldiers lacked the strength for prolonged effort, a failure for which the protein shortage in their diet may have been responsible. During World War I, as is well known, the undernutrition resulting from the Allied blockade contributed to the military defeat of Germany.

While wars provide the most spectacular and best-recorded examples of the influence of disease on history, other effects which are less well defined may have been of greater importance in the long run. By weakening the creative energy of people, disease can have cultural and political effects and thereby influence profoundly the fate of nations. Plague, for example, has upset the economic life of Europe on several occasions. In the sixth century A.D. it devastated the Roman world for fifty years and undoubtedly hastened its end. The Emperor Justinian himself was affected, but eventually recovered. Contemporary observers had been aware of the course of the epidemic. The Byzantine historian Procopius spoke of a plague "that spread over the entire earth, afflicting without mercy both sexes and every age. It began in Egypt, passed to Palestine and thence everywhere else."

The "Black Death" is said to have killed one fourth of the total European population during the fourteenth century, completely wiping out certain villages and towns. In England the repeated epidemics from the fourteenth to the seventeenth century culminated in the London outbreak of 1664–65. The account given by Daniel Defoe in his *History of the Plague*, in 1722, was largely imaginative, but subsequent historical research has brought out that the epidemics really had profound and varied effects on English social structure. For example, the shortage of labor resulting from the huge plague mortality compelled changes in agricultural practices. Grazing increased at the expense of cultivation, the tenant farmer was progressively replaced by the copyholder, and the enclosure movement spread over the land.

Malaria looms large from the point of view of its social effects. It was rife in earliest times in Babylon, Assyria, India, and southern China, and probably played a dominant role in the physical, intellectual, and moral decline of Greece. The ravages of malaria in the Roman Campagna were at times so great that they gave rise to the worship of *Dea febris,* the goddess of tertian and quartan fevers. The Campagna was a flourishing landscape four times in history: in pre-Roman time; when the Empire was at its height; in the eighth and ninth centuries; and during the Renaissance. In the intervals it was deserted because malaria made life impossible there. In brief, the Campagna was inhabitable whenever Roman hearts and muscles were robust enough to drain its marshes, whereas neglect of draining has always been associated with periods of economic decadence. It is worth recording in this regard the debt of gratitude that Europe owes to some of the monastic orders for eradicating malaria. There is no doubt that the disease was extremely prevalent in Europe before the continent was cleared of forests and marshes. When the Cistercians established their monasteries in the lowlands, at first in French Burgundy, they immediately engaged in extensive draining and, moreover, introduced cattle raising on their farms. Drainage decreased the breeding grounds for mosquitoes and those that survived fed on cattle by preference to man. The beneficial effects of this farming policy in controlling malaria became so well known that the Popes attracted the Cistercian order to the Campagna di Roma in the fifteenth century in the hope that they would duplicate in Italy the success that had attended their efforts in France.

Malaria has often hampered the development of new territories, even of certain southern regions of the United States. In Algeria the disease made the Mitidja Plain for a long time the tomb of settlers. Quite recently it has constituted a more difficult obstacle than the jungle and its creatures to the development of vast territories of northern India. Had it not been for his failure to control malaria and yellow fever, Ferdinand de Lesseps with French capital might have succeeded in opening the Panama Canal, with the consequence that the penetration of Central America by the United States would probably have been much delayed. It was because of its inability to deal with malaria and other tropical diseases that the white race failed to displace native populations of Africa during its great outburst of expansion in the nineteenth century. White men could take over and populate North America, South Africa, New Zealand, Australia, but they had to limit themselves to the role of financiers and supervisors wherever they could not control disease. Central Africa remained a black continent because it had become the white man's grave.

The almost universal infestation of hookworm in many lands of the South has had consequences greater than those of other factors more frequently discussed by historians and economists. The blood loss in human beings infested with hookworms long wasted the strength of these countries. Men lack physical vigor and initiative when they are made anemic through loss of blood or malnutrition.

In many parts of the world at the present time insufficiency of food is the one important factor which interferes with social creativeness. Studies carried out by the Germans during World War II provided the scientific demonstration, if any was needed, that the productivity of labor is related to the number of calories in the diet. No disciplinary measure, no threat or reward, could make men produce coal or steel when their strength failed through inadequate caloric intake. Future dictatorships, whatever their colors, will certainly use the science of nutrition to control the activities of men. The fight against hunger is the usual slogan of revolutionists, but in fact successful revolutions are never the feat of starving men.

As mentioned earlier, protein deficiencies are the worst aspect of food shortages in most of the world today. Plant proteins are deficient in certain amino acids, and few national economies can afford to produce meat or dairy products on an adequate scale. Furthermore, various other factors complicate still more the problems of meat production. The tsetse fly, which transmits trypanosomiasis to animals in certain parts of Central Africa, prevents the efficient raising of cattle. By depriving the native populations in these areas of meat and milk the fly contributes greatly to their intellectual and physical indolence. The religious taboos against killing of cattle have had similar effects in India.

Thus, diseases often interfere with the development of new agricultural practices and with industrialization. Yet the wealth needed for a large-scale attack on disease depends on these technological advances and it is not likely that the vicious circle can be broken without international co-operation. For this reason the success and very survival of the United Nations may depend eventually upon its ability to solve the complex problems of social technology involved in the relation of health to food supply.

Disease and Social Evolution

All groups of men, however primitive and loosely bound, have had some form of organization to protect communal health. Many taboos and religious customs certainly were built on a basis of pragmatic medical wisdom. The more human societies become complex and urbanized, the more they tend to regulate the behavior of their component members in an attempt to limit the damage caused by collective diseases. In the past, problems of health have provided compelling reasons for restricting certain individual freedoms, and as we shall see, some of their consequences are likely to accelerate the coming of the paternalistic state.

In Europe the hospital movement started with the "lazar houses," which were camps opened in the country for poor persons suffering from the disease of Lazarus (leprosy) or from plague or other contagious maladies. People moved by true charity, or wishing for particular distinction after death, or trying to be forgiven for some crime, took to visiting the lazar houses. By the end of the eleventh century leper visiting, like slumming in Victorian and Edwardian days in England, had grown quite fashionable despite the horrible stench of the lazarettos.

During the nineteenth century the knowledge that diseases hit particularly hard those in the poor economic brackets led many physicians to advocate social reforms. Furthermore, practical considerations soon convinced the general public, as well as industrialists and politicians, of the urgent need for concerted social action against the threat of disease. These various influences created a political climate in which preservation of the health of the masses came to be regarded as a responsibility of the community as a whole. "The health of the people," wrote Disraeli, "is really the foundation upon which all their happiness and all their powers as a state depend."

Elements of direct fear also contributed to the development of social medicine. The outbreaks of cholera had a prodigious effect. Eugène Sue's novel *The Wandering Jew* (1844–45) and Victor Hugo's poem *Chastisements* (1853) gave hair-raising accounts of the panic that they caused. In America cholera and, particularly, yellow fever in 1878 stimulated the creation of a national board of health and then of special laboratories supported by public funds for the control of water supplies. This step led to the granting of ever-increasing power to health departments for the regulation of community life. Another phase of the socialization of medicine was ushered in by fear of tuberculosis. It was a short step from the demonstration that the tuberculous individual could infect his fellow men, and that tuberculosis was therefore a social disease, to the use of public funds for the control of the disease and even for the care of the patient. The trend toward socialization of medicine is still continuing, although there is reluctance in designating the process by this name. In one form or another, many aspects of communal activity are regulated, restricted, or prevented because of their effects on public health. Strict regulations will certainly extend to the control of industrial smokes and exhausts of motorcars, as soon as the public becomes emotionally convinced that these nuisances constitute health hazards. All modern states, whatever their political complexion, recognize that the maintenance of health is as much a government responsibility as is education.

While the fear of disease has been indirectly the cause of legislation limiting personal freedom, the prolongation of life in Western countries is now encouraging the growth of the welfare state. As the percentage of individuals past the wage-earning age becomes larger, there is increasing demand for social security and for a planned economy to provide for old age. The political philosophy of a population made up largely of young and adult men eager for economic expansion is bound to differ from that of a community consisting in large part of older individuals concerned with problems of retirement.

The fall in death rate during childhood also is likely to exert a profound influence on the social structure and perhaps on the whole future of the human race. Until a few decades ago a large percentage of children died of infection during the first years of their life. Queen Anne probably holds the record among the famous of the world, with seventeen pregnancies and not a single survivor. Now most of the children born in Europe and North America survive, a fact which has had a dominant influence in bringing about planned parenthood. This control of population level facilitates the securing of a stable population enjoying a decent standard of living, but it also decreases the willingness to take risks and increases the demands for security. Furthermore, it may have unforeseen biological consequences. Now that techniques have been worked out to permit survival of all children, and social, religious, and medical ethics demand that all be allowed to live and reproduce their kind, humanity faces a state of affairs which is without precedent in the biological world and which bids fair to present new problems to medical and social sciences in a not too distant future.

Effects of Disease on Cultural Forces

Much has been written concerning the influence that the state of health of a particular individual has had on his artistic creation. R. L. Stevenson spoke of the deprivation which his recovery from tuberculosis meant in loss of stimulus to his artistic faculties. Asthma obviously played a large role in Marcel Proust's perception of the world. Of greater interest, however, are the relationships between art and disease, which concern society as a whole. Some of these are manifested so directly in the various forms of plastic expression that much useful knowledge concerning the prevalence of diseases can be derived from pictorial and sculptured records. For this reason, pottery, bas-relief, statuary, and painting provide useful documents for the medical history of ancient times Charcot assembled two remarkable collections, *Les Démoniaques dans l'Art* and *Les Difformés et les Malades dans l'Art,* which convinced him that hysteria was not a new disease as was commonly thought in his time, but had affected men as well as women in all ages. *Plague and Pestilence in Literature and Art,* published by Raymond Crawfurd in 1914, dealt with a theme that Rudolf Virchow himself had inaugurated half a century before. According to Virchow, "A portrait of the holy Elizabeth painted in 1516 by Holbein the Younger which I discovered in the Munich gallery illustrates leprosy as it existed in Germany, as well as syphilis. Renewal of interest in old paintings from various countries has helped to identify leprosy of former days more satisfactorily than is possible from bare descriptions."

The prevalence of leprosy and of plague, as well as the fear that they inspired, are reflected in countless illustrations and paintings of the Middle Ages and the Renaissance. Because St. Sebastian was claimed to have powers of protection against plague, Georges de La Tour painted many portraits of him for his wealthy patrons.

Tuberculosis was a common disease in the urbanized society of the Italian Renaissance. Simonetta Catanea Vespucci (1459–75), a Florentine beauty who died of consumption at an early age, had been Queen of Beauty in a tournament and had won the admiration of Lorenzo de' Medici. She sat on many occasions for Botticelli, and her type continued to be used by painters even after her death. The stigma of disease can be seen in the sunken cheeks, long slender neck, steep sloping shoulders of Botticelli's models. Two other young consumptives, the respective wives of Dante Gabriel Rossetti and William Morris, served as models for the "long, cadaverous women with sensuous lips," as G. B. Shaw saw them, so commonly portrayed by the English Pre-Raphaelite school of painting.

Oddly enough, these elongated emaciated women and their chlorotic sisters shared the stage in nineteenth-century pictorial art with the rotund bourgeois fond of port and roast beef, whose gouty toes appear so frequently in Hogarth and Rowlandson's caricatures. Rowlandson united two of the largest medical problems of the nineteenth century in his satirical print "Dropsy Courting Consumption." And Honoré Daumier amused himself by showing in one of his lithographs a corpulent hypochondriac of the middle class confessing with anguish to the physician who is examining him, "Oh, Doctor, I'm sure I'm consumptive."

It is not only by providing pathological human types as models that disease influences the manifestations of art. Plague, the Black Death, had a more subtle but nevertheless profound influence on the history of the Renaissance through many other indirect effects. Immediately after 1350 dire prophecies of new pestilences to come prompted a revival of Catholic Christianity with the donation of huge sums of money for the creation and embellishment of churches, chapels, and monasteries. Saintly personages appeared, several of whom were subsequently canonized. The immense death toll in Florence forced relaxation of the guild laws which limited the immigration of artisans, physicians, and jurists into the city. Within a few years a class of nouveaux riches thus arose out of the reshuffle of wealth and their unsophisticated tastes imposed more conservative formulas of creation upon the artists whom they commissioned.

Directly or indirectly, the Black Death appears frequently in the literature of the Renaissance. For Petrarch it had been a great personal calamity. The deaths of Laura and four of his closest friends contributed to turning him to religion. The ten young men and women of Florence assembled by Boccaccio in the *Decameron* had fled to the hills in an attempt to escape from plague. It was the destructive epidemic of 1348 which had brought the raconteurs to the villa where they entertained themselves with the tales which have entertained the world ever since. Soon, however, Boccaccio himself lost his mirth under the influence of the dark state of mind engendered in all social classes by the Black Death. The pestilence was almost universally regarded by his contemporaries as the manifestation of the wrath of an aroused God punishing mankind for its wickedness. The sense of guilt and the resultant asceticism are reflected in Boccaccio's novel, *Il Corbaccio*, published in 1354. In fact, the gloomy vision of the world which then pervaded most of Italy eventually led the author of the *Decameron* to turn bitterly against his own early work.

Plague also gave a poignant meaning to the story of Job, an old paradigm of trial and affliction. Because Job suffered from a disease with symptoms outwardly similar to those of plague, he came to be often portrayed in Tuscan paintings. In Italian and northern art of the fifteenth century the conception of an aroused God punishing mankind by pestilence often assumed the form of Christ hurling arrows at the world, like the thunderbolts of Jupiter. Localized outbreaks of plague continued in Europe until the seventeenth century, inspiring fanatical devotion and asceticism in certain groups and loosening all social and moral restraints in others. Plague and its social effects in seventeenth-century Italy gave to Alessandro Manzoni the subject of a famous passage in *I Promessi Sposi* (*The Betrothed*), the first novel of importance in modern Italian (1825).

Venereal diseases have exerted a complex influence on

the social fabric of the Western world, the ultimate mani-festations of which are yet to unfold. There has been much argument regarding the time of appearance of gonorrhea and syphilis in Europe. Most historians of medicine believe that gonorrhea is a very ancient disease, referred to in the Bible and described by the Greeks and Romans. It was recently pointed out by the English physician H. St.H. Vertue, however, that the disease described by Galen and other ancient writers was spermatorrhea and not the con-tagious urethritis caused by the gonococcus. Moreover, still according to Dr. Vertue, the type of social relation between sexes and the sexual mores prevailing in the Greco-Roman world were hardly compatible with a widespread distribu-tion of *lues venerea*. In fact, none of the amatory writers of Rome ever alluded to the existence of venereal diseases, not even Propertius in his poems, Ovid in his *Ars Amatoria,* Horace in his *Epodes,* Juvenal in his satires, Martial in his epigrams, Plautus or Terence in their comedies. Venereal diseases are not mentioned either in *The Arabian Nights,* in Villon's poems, in Boccaccio's *Decameron,* in Chaucer, Langland, Gower, or any of the other English medieval poets. It is not likely that all these writers, otherwise so uninhibited in the selection of topics and in the description of all kinds of experience, would have refrained from men-tioning gonorrhea or syphilis if these venereal diseases had been of common occurrence and constituted a danger in the societies of their times. The absence of venereal dis-eases, Dr. Vertue claims, accounts in part for the casual attitude of the ancient and medieval worlds toward sexual problems.

The first unquestionable reference to gonorrhea appears in a manuscript by John of Ardenne published in England in 1378. From the fifteenth century on, gonorrhea is freely and frequently mentioned in European literature, often un-der the name "burning" or "clap." A medical description was presented by Andrew Boord in 1546, and the disease appeared even in Shakespeare. In the meantime, however, a more terrible malady had appeared on the scene. The first well-recognized outbreak of syphilis occurred in Naples in

February 1495. As Christopher Columbus returned from Hispaniola on March 4, 1493, the conjunction of the two events naturally suggested that syphilis had been contracted by the Europeans from the Amerinds. Although this conclusion is still *sub judice*, it is clear that if Europeans knew syphilis before 1495 they troubled themselves little about it. In fact, the physicians as well as the public were certainly surprised by the disease and, having no name for it, chose to trace its origin to those whom they made responsible for its spread. The French referred to the *"maladie napolitaine"* and the Italians to the *"maladia francese."* Only later did syphilis come to be known as the Indian disease. The Rouen physician Jacques de Béthencourt referred to it as a "venereal" disease in 1527. What is certain is that syphilis soon became a major problem among the French troops during the Italian campaigns and in fact was partially responsible for their ultimate failure. It provided the subject for the last important literary work in the Latin language, the poem *Syphilis sive Morbus Gallicus*, published in Verona by Fracastoro in 1530.

During a few generations syphilis spread with great virulence throughout most of Europe, affecting all aspects of human life. Because it frequently caused a loss of hair in its victims it brought about the general use of the ruff and the wigs as a part of masculine costume. Public baths, which had been a feature of European life since Roman times, came to be regarded as a source of contamination and for this reason were rapidly abandoned.

Rightly or wrongly, syphilis has been made responsible for many of the ills of the modern world. The terrible outbreak of 1495–1520 occurred at one of the peaks of European history, when the Renaissance had reached its greatest heights in art and learning, daring and industry. After syphilis, superstition and gloom and often panic badly tarnished the brilliance of European civilization. Religious sermons pointed to the spread of venereal diseases as evidence of God's anger at mankind and as a proof that sex was an ally of sin. All forms of sexual offense, and in particular adultery, came to be looked upon with greater severity. It is doubtful that religious sermons were then much more effective than they are now in promoting sexual restraint, but it is probable that the fear of venereal diseases often constituted a deterrent to promiscuous relations. Samuel Pepys provides in his diary the candid revelation of an attitude widely held but not often expressed. In the entry for December 14, 1762, he writes: "Have been in London several weeks without ever enjoying the delightful sex. . . . Many fold are the reasons for this my present wonderful continence. . . . I have suffered severely from the loathsome distemper, and therefore shudder at the thoughts of running any risk of having it again. Besides, the surgeons' fees in this city come very high." As late as 1902 it was stated in a German magazine that if syphilis, "the punishment inflicted by nature on vicious men," should ever become curable, society and morality would greatly suffer from "a moral syphilisation even worse than that of the body."

Gonorrhea and syphilis are now among the diseases most readily amenable to drug treatment, for example, with penicillin, and it can be expected therefore that the fear of disease will not much longer constitute a deterrent to promiscuity. On the other hand, it is obvious that many biological and social factors other than fear of disease also influence sexual mores. Had Kinsey-type surveys been available for the preceding centuries, they might have revealed useful information with regard to the effects brought about by the penicillin era in the relationships between the sexes. In the absence of precise sociological studies, the subject matter of future plays and novels may indicate to what extent ethics, aesthetics, and sentiments can act as substitutes for the "loathsome distemper" in regulating the sexual behavior of man.

During the nineteenth century the growth of social conscience added the sense of responsibility to the fear of syphilis. Ibsen dealt with this aspect of the problem in his plays *Ghosts* (1881), *An Enemy of the People* (1882), and *A Doll's House* (1879). In 1901 Eugène Brieux's play *Les Avariés,* translated into English as *Damaged Goods,* caused a storm in Europe and focused public attention on what had come to be called *the* social evil. The theme was further exploited in other novels and plays, such as Le Gouradiec's *The Fatal Kiss* (1918). The association of sin with sex reached its most intellectual expression in Thomas Mann's *Doctor Faustus* (1947). In the book the hero Adrian is talked into the belief that "The act of procreation, esthetically disgusting, is the expression and the vehicle of original sin" and deliberately embraces a prostitute who had warned him she was syphilitic.

D. H. Lawrence was convinced that "the effects of syphilis and the conscious realization of its consequences" had done more than any other factor to alter the atmosphere in Europe after the sixteenth century. According to him, it gave a "great blow to the Spanish psyche" and conditioned the puritan attitude in England and America. The word "pox" was in every mind and in every mouth among the Elizabethans, and "Pox on you" was the common oath.

It is one of the words that haunt Elizabethan speech [wrote D. H Lawrence]. The secret awareness of syphilis, and the utter secret terror and horror of it, has had an enormous and incalculable effect on the English consciousness and on the American. Even when the fear has never been formulated, there it has lain, potent and overmastering. . . . I am convinced that *some* of Shakespeare's horror and despair, in his tragedies, arose from the shock of his consciousness of syphilis. *Some* of Shakespeare's father-murder complex, *some* of Hamlet's horror of his mother, of his uncle, of all old men came from the feeling that fathers may transmit syphilis, or syphilis-consequences, to children. . . .

The terror-horror element which had entered the imagination with regard to the sexual and procreative act was at least partly responsible for the rise of Puritanism, the beheading of the king-father Charles, and the establishment of the New England colonies. If America really sent us syphilis, she got back the full recoil of the horror of it, in her puritanism. . . .

With the collapse of the feeling of physical, flesh-and-blood kinship, and the substitution of our ideal, social or political oneness, came the failing of our intuitive awareness, and the great unease, the *nervousness* of mankind. We are *afraid* of the instincts. . . .

Now we know one another only as ideal or social or political entities, fleshless, bloodless, and cold, like Bernard Shaw's creatures.[1]

During the nineteenth century tuberculosis was the greatest single cause of disease and death. Because it killed so many young men and women, and wounded so many hearts, it contributed certainly to the melancholy mood of the romantic era by introducing in daily experience the stabbing sense of the brevity of life. This was the time when, in Keats' words, "Youth grows pale, spectre thin and dies." The fading away of young women dying of consumption—in a decline, as it was then proper to say—became a poetical theme of literature. The heroines of Edgar Allan Poe's stories and poems were modeled after his young wife, Virginia, dead of tuberculosis at the age of twenty-four. The tragic atmosphere pervading the novels of the Bronte sisters reflects the prevalence of disease all around them. The four sisters and their brother died consumptive in their youth or early adulthood.[2]

In French literature Marguerite Gautier of *La Dame aux Camélias* or *La Traviata*, the pathetic Mimi of *La Bohême*, both of whom were made to die of tuberculosis on the stage, were not entirely fictional characters. Their prototypes had been the mistresses of Alexandre Dumas *fils* and Henri Murger and their true history is known in great detail. Like so many young women of their generation, they were consumptive and died in their early twenties, in 1847

[1] Quoted from D. H. Lawrence, *Selected Literary Criticism* (ed. by Anthony Beal). New York: Viking Press, 1956.

[2] For further details on the social and cultural manifestations of tuberculosis, see *The White Plague—Tuberculosis, Man and Society*, René and Jean Dubos. Boston. Little, Brown and Co., 1952.

and 1848, respectively. In real life Marguerite Gautier's name had been Alphonsine (Marie) Duplessis (1824–47). After her death Dickens attended the sale of her belongings and reported that "To see the general admiration and sadness, one could have believed that it had to do with a Jeanne d'Arc." Dumas wrote *La Dame aux Camélias* under the influence of the news of her death. Among the many other young tuberculous women who provided inspiration for the French romanticism were Pauline de Beaumont for Chateaubriand and Julie Charles for Lamartine. Nicolò Paganini, "pallid and corpse like," Rachel, the divine actress, "white like alabaster," were but two of the artists whose dramatic effectiveness on the stage was heightened by tuberculosis.

Even literary symbols and images reflect the prevalence of tuberculosis in the nineteenth century. To the poetic souls of the period, autumn did not signify the time of crops and abundance but the death of everything in nature. The dying foliage symbolized the fate of the consumptive. In "Ode to the West Wind," Shelley described the falling leaves as "pale and hectic red," like "Pestilence-striken multitudes." And Thoreau, seeing the first spotted maple leaves "with a greenish center and a crimson border," remarked in his *Journal* that "Decay and disease are often beautiful, like . . . the hectic glow of consumption."

So prevalent was the atmosphere of disease at that time that good health came to be considered almost a sign of vulgar taste. Certain forms of disease, consumption in particular, were thought to endow their victims with a peculiar quality of spirituality, even with creative genius. Like many others, the Goncourt brothers often evoked with admiration in their *Journal* the fascinating appearance of a sick woman: "I thought how dangerous it would be to meet this woman too often, a danger composed of the immateriality of her person, the supernatural character of her glance, the emaciation of those features of an almost psychic fineness, that something suprahuman as if belonging to one of Poe's heroines become a Parisian." The Russian wonder child Marie Bashkirtsev also accepted disease as contributing to her charm and noted in her diary: "I cough continually! But for a wonder, far from making me look ugly, this gives me an air of languor that is very becoming."

The wasting and emaciation caused by phthisis added to the glamour of many of the romantic artists and poets just as languor did to the charm of young women. "I should like to die of a consumption," once said Byron, "because the ladies would all say, 'Look at that poor Byron, how interesting he looks in dying!'" Sidney Lanier, pale, dark, slender, and nervous, also a victim of tuberculosis, was the ideal of the bard in the United States. It was consistent with the general attitude that disease and society had thrust upon him that, despite his admiration for Walt Whitman, he was shocked by the healthy animality of the poet's genius. Sainte-Beuve, who grew to be the fat bon vivant shown by his later portraits, nevertheless started his literary life by depicting himself under the name of Joseph Delorme as a pale medical student, suffering from pulmonary phthisis. Even the robust and sensuous Alexandre Dumas made occasional attempts to look frail and consumptive. "In 1823 and 1824," he writes in his memoirs, "it was the fashion to suffer from the lungs; everybody was consumptive, poets especially; it was good form to spit blood after each emotion that was at all sensational, and to die before reaching the age of thirty."

The attitude of perverted sentimentalism toward disease continued almost to the end of the nineteenth century, although it took slightly different forms to adapt itself to the changes in prevalence of types of diseases. The episode of the last leaf of autumn in La Bohème refers to tuberculosis. In O. Henry's story The Last Leaf the heroine dies of pneumonia, a pulmonary disease that was then the great killer in New York City. By the turn of the century, however, the romantic view of disease was becoming less fashionable. Thomas Mann's Magic Mountain was the last masterpiece inspired by tuberculosis and its approach was more intellectual than romantic. The disease was changing, as was the social attitude toward it. Referring to sanatorium life, Mann wrote thirty years after creating his novel: "The

Magic Mountain became the swan song of that form of existence. Perhaps it is a general rule that epics descriptive of some particular phase of life tend to appear as it nears its end."

Even before this final phase the revelations of the public-minded sanitarians had opened the eyes of writers and artists. They discovered that disease was rarely poetical and that the artificialities of the romantic era gave a very distorted picture of the state of mankind in the Western world. Instead of glamorizing the fainting young women and their romantic lovers, writers took notice of the miserable humanity living in the dreary tenements born of the Industrial Revolution. In the "tentacular cities" they saw hosts of men, women, and children, pale from hunger and cold, working long hours in dark and crowded shops, breathing smoke and coal dust. Disease was there, but its expression was physiological misery and suffering without romance. It dawned on social common sense what a mockery it was to depict disease as a spiritualization of the being Consumption was not an aristocratic decline leading to an ethereal release of the soul among the falling autumn leaves, it was instead the great killer and breeder of destitution Influenced by the objective knowledge derived from modern medicine, the writers of the realist school saw disease not as a romantic experience but in the form of emaciated bodies, of faces livid and distorted by cough, of minds haunted by the fear of poverty and death. Balzac had voiced this change in the preface to *La Peau de Chagrin*. He scorned "the *infirmerie littéraire*" and stated his conviction that the public was tired of reading about the "sad, the leprous, the langorous elegies." Théophile Gautier also acknowledged that after having been a "thorough bred romantic" convinced that no "lyric poet could weigh more than 99 pounds," he had passed to the opinion that "the man of genius must be obese."

The germ theory caused a still further evolution among writers and the general public. Disease had been exploited by the Victorian novelist as the manifestation of an inexorable fate that won the sympathy of the reader for the hero Disease now became a contagion, something unclean. The infected individual in the modern novel is almost an untouchable, at times arousing repulsion or fear. It was this new attitude which found expression in Mauriac's *Le Baiser au Lépreux* and Julian Green's *Adrienne Mésurat*.

Samuel Butler's *Erewhon*, published in 1872, was in part also a form of reaction against the romantic glorification of suffering. In Butler's utopia, illness was considered a crime. For the trial of a consumptive described in the novel, Butler took the judge's summing up from a newspaper report of the case of a man found guilty of theft, with scarcely more alteration than the name of the offense. "Whether your being in a consumption is your fault or not, it is a fault in you . . . you may say that it is your misfortune to be criminal; I answer that it is your crime to be unfortunate." In *Erewhon* the wayside shrines were statues of men and women in full flower of their strength, youth, and beauty.

Just as diseases change in prevalence and severity, so do their effects on literature and on the arts The young woman who was made attractive or pathetic two generations ago by languishing into a romantic decline now glories in physical vigor and a tanned complexion. A hundred years ago the typical hero of the novel suffered and died of some disease of languor, preferably consumption, unless he became victim of some more sudden plague, like cholera or yellow fever. The novelist of today makes use of more modern weapons to eliminate or incapacitate his characters.

As yet he has hardly ever brought cancer into play. The very word still conveys so much awe that it is unspeakable, as once were syphilis and tuberculosis. There are other forms of disease, however, that the writer can use to create the type of gloom characteristic of our age. High blood pressure, heart disease, cerebral hemorrhage, gastric ulcer, and acts of violence conveniently and convincingly jeopardize or destroy the dominant characters of present-day fiction. Their fate is often the penalty of ambition and social success, as appears from Cameron Hawley's novel *Executive Suite*, which takes as its object business life in modern

America. The plot is introduced by the death of a prominent industrialist from cerebral hemorrhage at the age of fifty-five. His predecessor had shot himself some ten years earlier and one of his competitors suffers a stroke later in the book. As to the wives and daughters of these tycoons of industry and finance, they do not fare better: one commits suicide and another must be hospitalized in an institution for the mentally ill. Peptic ulcers seem to be an obsession. The manager of the New York office knows that the apprehension of seeing his boss appear unexpectedly is the cause of his gastric ulcer. Arrangements are made to supplement the salary of an executive by giving him the use of a private airplane, but a person wise in the ways of the world remarks that "he would not trade a duodenal ulcer for a DC3." "To become president of a corporation," affirms a member of the board later, "is one of the least rewarding forms of suicide."

This last statement reflects an attitude now sufficiently prevalent to constitute an important social problem. The belief that success in a highly competitive society is often bought at the cost of health and happiness leads men, even young men, to spurn adventure for the sake of security and comfort, even though it implies mediocrity. The pioneer is still a hero on the screen, but the viewer elects for himself the life of the suburban petit bourgeois. There is some evidence in fact that a similar change of heart has happened more than once in history.

Sometime around 600 B.C. there spread through China the Taoist philosophy, which was probably a social reaction against the disorders of the chaotic era known as the Period of the Six Dynasties. The founder of Taoism was the poet-philosopher Lao-tzu, who had been an important and successful official at the Imperial Court. Disheartened by the endless strife and the futility of life in the China of his time, Lao-tzu decided to retire to the peace and quiet of the mountains. As he was passing through the gates of the Great Wall he was recognized by minor officials, who urged him to formulate his philosophy before withdrawing from the world. This Lao-tzu did in a poem known as *Tao*

Tê Ching, or *The Way of Life* Tao urges man to identify himself with the natural world around him, to merge his activities, emotions, and sensations with the rhythm of seasons and the processes of nature, and in particular to lead a life devoid of useless strife and struggles Tao, The Way, is the road not only to happiness but also to health. According to Lao-tzu,

> *Those who flow as life flows*
> *Feel no wear, feel no tear*
> *Need no mending, no repair.*

For centuries Lao-tzu's poem has inspired many phases of Chinese life, contributing to it a certain type of submissiveness. The haunting serenity of the paintings of the Sung era is the immortal expression of Taoist philosophy and of its ideal of identification of man with nature. It is the same ideal of a society without conflicts that was in the mind of the Yellow Emperor, evoking in his *Classic of Internal Medicine* the days when men could live to a hundred years and be spared illness by avoiding strife and following the laws of the seasons.

In our society at the present time it is not by living in harmony with nature and in accordance with the laws of the seasons that man hopes to find health, comfort, and security. The welfare state and its medical counterpart are now expected to bring the solutions to all these problems. And, indeed, it is legitimate to hope that with enough regulations, restrictions, and injections man could achieve effective control over most of his fatal diseases. But too often the goal of the planners is a universal gray state of health corresponding to absence of disease rather than to a positive attribute conducive to joyful and creative living. This kind of health will not rule out and may even generate another form of ill, the boredom which is the penalty of a formula of life where nothing is left unforeseen.

In appearance the artist is far removed from these problems. Whatever technique he uses, he is wont to be interested more in form and expression than in subject matter. But even when he pretends to be concerned with art only for art's sake he cannot escape the impact of the world in which he lives. Even abstract art is to a large extent a protest against the pressure of an objectionable reality. Today, as in the past, problems of health and disease are among the factors of the environment which influence the topics and moods of artistic creation. We may fail to identify this influence, just as the Middle Ages and the nineteenth century were unaware of the fact that hunger, malnutrition, filth, and pestilence colored their attitudes and emotions. But it is certain that the pathological manifestations of our competitive and mechanical life are reflected in our cultural climate. The diseases of our time, from hypertension to silent despair or acute paranoia, cannot help imparting to the music, the written word, and the plastic arts of the present era a mood which to the critic of the future will seem as weird and unhealthy as the sentimental excesses of the romantic age appear to us today.

Directly or indirectly, the various forms of art reflect the strivings, the struggles, and the sufferings of mankind. The state of health and the ills of a society are recorded not only in the writings of its physicians and scholars but also in the themes and moods of its artists and poets.

VIII.

UTOPIAS AND HUMAN GOALS

Arcadias and Utopias

The ancient province of Arcady lies in the heart of the Peloponnesus, all but isolated from the rest of Greece by mountains. In the legend it was the domain of Pan, who played the syrinx on Mount Maenalus, and of rustic people celebrated for their musical accomplishments and their rustic hospitality, but also notorious for their ignorance and low standards of living. Yet it was this unfavored land, poor, rocky, chilly, devoid of all the amenities of life, affording adequate food only to goats, which was transformed through the alchemy of art into the myth of Arcadia. From Vergil to Nicolas Poussin, "I, too, dwelt in Arcady" has symbolized the golden ages of plenty and innocence, of unsurpassable happiness enjoyed in the past and enduringly alive in memory.

While the Greco-Roman civilization placed its land of dreams in a remote and not easily accessible Arcadia, Chinese Taoism found it in any place where man could achieve identification with nature—in romantic mountain paths, isolated fishing villages, or mist-bathed landscapes. According to Lao-tzu and his Taoist followers, joy and bliss were possible only in a world of primitive simplicity. Men could achieve health and happiness only by merging themselves with their environment and living in accord with the laws of the four seasons, by participating with other living creatures "in the mysterious equality and thus forget themselves in the Tao."

The Taoist's withdrawal from conflict and his attempt to identify himself with the physical and social environment constituted a philosophy of health. Avoidance of travel minimized the transfer of new pathogens from one community to another. Life without aggressive behavior and in accordance with the rhythms of the seasons made it possible to reach a state of harmony with the environment. This way of life was not designed to solve the difficulties arising from social contacts and conflicts. Rather, it attempted to prevent or at least to minimize the emergence of new problems by creating a stable world in which new stresses, but also new experiences, were ruled out.

While the Arcadian bliss and the contented intimacy of the Chinese Tao are rarely attainable in real life, they constitute eternally the stuff of human dreams. As a substitute for the Arcadias of the past, men never tire of imagining for the future new types of social order free of the defects and vices found in all actual societies. But utopias differ profoundly one from the other despite their common basis of illusion, because each is colored by the value judgments of its originator. Utopian ideals vary all the way from a desire for nirvana to the longing for exciting experience; from the passivity, indolence, and tolerance of Goncharov's oblomovism to the ceaseless activity and creative endeavor of the Faustian universe.

Propounders of utopias have not even been able to agree on the value that they attach to life. Plato considered that life without health was not worth preserving for the sake of either the individual or the community. He saw no virtue in encouraging the survival of a fellow man threatened by continuous sickness. The state physicians of his *Republic* were to watch with care over "the citizens of goodly conditions, both in mind and body," but persons who were defective either mentally or physically were "to be suffered to die." This attitude is a far cry from the ethics of modern utopias. Life, it is now taught, must be preserved at all cost, whatever the burden that its preservation imposes on the community and on the individual concerned. Whether this lofty ethical concept will retain acceptance if put to the acid test of social pressure still has to be proved. Western man may rediscover wisdom in Plato's social philosophy when the world becomes crowded with aged, invalid, and defective people. He may once more rationalize himself into the belief that happiness is not possible in the absence of usefulness to the social group and that survival under these conditions is therefore not worth having.

Designers of utopias must also formulate judgments of value regarding the type of human beings they want to foster. The society best suited for producing athletes, warriors, and men of action is not necessarily the best breeding ground for artists, scholars, philosophers, and mystics. In addition, many trivial factors, conscious or unconscious, influence the community in determining the defects that it will tolerate and the level of physical and intellectual adequacy to which it aspires. Most Western societies today regard as unacceptable certain smells or skin blemishes which were a matter of course a few generations ago and are still accepted as the normal state by many primitive or semi-civilized peoples. Modern man looks with dismay on the fact that syphilis, malaria, yaws, intestinal disorders, etc., are so common in some areas of the world as not to be regarded as diseases. Yet he accepts as part and parcel of a normal life baldness, poor eyesight, chronic sinusitis, and other bodily defects which might be regarded as handicaps or even as repulsive traits in other cultural contexts.

Clearly, health and disease cannot be defined merely in terms of anatomical, physiological, or mental attributes. Their real measure is the ability of the individual to function in a manner acceptable to himself and to the group of which he is a part. If the medical services of the armed forces seem more successful than their civilian counterparts in formulating useful criteria of health, this is due not to their greater wisdom but rather to the fact that their criteria are more clearly defined. On the whole, effective military performance required attributes less varied and less complex than the multifarious activities of civilian life. But criteria of adequacy change even in the military world. The soldier of past wars who marched or rode his way to victory through physical and mental stamina might not be the most effective warrior in the push-button operations of future conflicts.

For several centuries the Western world has pretended to find a unifying concept of health in the Greek ideal of a proper balance between body and mind. But in reality this ideal is more and more difficult to convert into practice. Poets, philosophers, and creative scientists are rarely found among Olympic laureates. It is not easy to discover a formula of health broad enough to fit Voltaire and Jack Dempsey, to encompass the requirements of a stevedore, a New York City bus driver, and a contemplative monk.

One of the criteria of health most widely accepted at the present time is that children should grow as large and as fast as possible. But is size such a desirable attribute? Is the bigger child happier? will he live longer? does he perceive with greater acuity the loveliness or the grandeur of the world? will he contribute more to man's cultural heritage? or does his larger size merely mean that he will need a larger motorcar, become a larger soldier, and in his turn beget still larger children? The criteria of growth developed for the production of market pigs would hardly be adequate for animals feeding on acorns in the forests and fending for themselves as free individuals. Nor are they for man. Size and weight are not desirable in themselves, and their relation to health and happiness is at most obscure. In his essay *On the Sizes of Things or the Advantages of Being Rather Small,* Boycott concluded, in fact, that an animal about as big as a medium dog has the best possible size for our world!

Curiously enough, the assumption that human beings should grow fast and large has never been examined closely as to its validity and ultimate consequences. Its only certain merit is that weight, size, and a few other physical traits can be measured readily, provide objective and convenient characteristics on which to agree, and can be on the whole readily achieved. There is no evidence, however, that these

criteria have much bearing on happiness, on the development of civilization, or even on the individual's ability to
adapt to the complex demands of modern technology.
While high humidity usually enhances the development of
orchid plants, it is not particularly favorable to the development of the flowers, *Grevillea robusta,* which provides valuable timber under the relative drought conditions of Australia, yields but valueless wood when caused to grow
rapidly as a shading plant on the coffee plantations of the
tropical Guatemalan highlands. For man, similarly, mere
size has never been the determinant factor of his survival
and success, either as an individual or as a species. Large
size is likely to prove even less of an asset in the world of
the future, and may even become a handicap. The specifi-
cations for man's body and mind may have to be reformulated in order to meet with greater effectiveness the exigencies of the mechanized world.

Arcadias are dreams of an imaginary past, and utopias
the intellectual concepts of an idealized society. Different
as they appear to be, both imply a static view of the world
which is incompatible with reality, for the human condition has always been to move on. "Man has never sought
tranquillity alone," wrote Sir Winston Churchill. "His nature
drives him forward to fortunes which, for better or for
worse, are different from those which it is in his power to
pause and enjoy." Prehistory and ancient history show that
men have never been able to forget their nomadic past and
to rest quietly in the corner of the earth they had made
their own for a while. Not satisfied with changing their
geographical environment, men also crave for changes in
their social atmosphere. Their utopias have never been able
to keep pace with their fundamental restlessness, with their
eternal dream of a New Jerusalem.

From Biological Adaptation to Social Evolution

Fossil remnants of prehistoric man have been found in
greatest profusion in East Africa. It seems that Ethiopia,

Kenya, Tanganyika, and neighboring countries have provided conditions well suited for the evolutionary changes through which the human race achieved the diversity which permitted it to colonize the whole world. On the one hand, much of East Africa consists of highlands with a moderate climate varied enough to produce the stimuli required for the evolution of an all-purpose ancestor of man. On the other hand, this region offers a large variety of geological strata, topographical configurations, climates, fauna, and flora to which early man could gain easy access. Within a few hundred miles are to be found high peaks, rich plateaus and lands below sea level; torrential waters, immense lakes, and gentle seas; tropical forests, alluvial plains, and deserts of sand. Thus, even short migrations provided for man in this area the opportunity to gain experience with and achieve biological fitness to a wide range of physical environments. And he did not have to travel far to reach the lands where he was to embark on his cultural destiny. From the Abyssinian mountains, the Blue Nile opened for him a channel to the luminous and fertile deltas of the Near East which became the cradles of his civilizations.

As he moved into new lands and new climes man underwent adaptive biological changes in response to the various environments that he encountered. To a large extent this biological phase of evolutionary history seems to have been completed by the end of the Pleistocene epoch. Physical man was then essentially a *fait accompli*. But, while the size of his brain, his physiological reactions, and even his fundamental instincts have probably changed little since that time, the social structures that he has developed have continued to evolve. It is clear that the collective evolutionary course of mankind has now set the human species apart from the rest of the animal world. The present phase of human evolution differs qualitatively from the purely biological phase because passive submission to the environment has been replaced by an active creative process. Evolutionary changes which were once the slavish expressions of natural forces have become increasingly self-directing. They affect not so much the body and the mind of man as the type of life that his social organization makes possible. Their effectiveness is based on the ability to acquire and transmit information in a manner that gives to the social body the cumulative experience and knowledge of each of its members.

All these new aspects of human activities are identified with the invention of tools and the development of social groups. Communal life, in villages and then in cities, created new environmental problems that stimulated new adaptive processes. This major change occurred only some ten thousand years ago. At the rate of three generations per century, this lapse of time is far too short to have allowed adequate play for the usual mechanisms of biological adaptation. Rather, it was through the development of social practices that man met the countless and unexpected new challenges that he encountered in the course of his migrations and social upheavals. Religious beliefs, empirical wisdom, and eventually scientific understanding played dominant roles in helping him to resist threats originating from nature or, more often, from his own activities. Whereas other living things survive through adaptive changes in their bodies and their instincts, man strives to impose his own directional will on the relations that he has with the rest of the world. Consciously, though often not wisely, he decides on the kind of life he wishes to have; then he acts to render possible this way of life by shaping the environment and even attempting to alter his own physical and mental self.

Social Changes and Ecological Equilibria

Modern man believes that he has achieved almost complete mastery over the natural forces which molded his evolution in the past and that he can now control his own biological and cultural destiny. But this may be an illusion. Like all other living things, he is part of an immensely complex ecological system and is bound to all its components by innumerable links. Moreover, as we have seen, human life is affected not only by the environmental forces presently at work in nature but even more perhaps by the past.

Any attempt to shape the world and modify human personality in order to create a self-chosen pattern of life involves many unknown consequences. Human destiny is bound to remain a gamble, because at some unpredictable time and in some unforeseeable manner nature will strike back. The multiplicity of determinants which affect biological systems limits the power of the experimental method to predict their trends and behavior. Experimentation necessarily involves a choice in the factors brought to bear on the phenomena under study. Ideally, the experimenter works in a closed system, affected only by the determinants that he has introduced, under the conditions that he has selected. Naturally, however, events never occur in a closed system. They are determined and modified by circumstances and forces that cannot be foreseen, let alone controlled. In part this is because natural situations are so complex that no experimental study can ever encompass and reproduce all the relevant factors of the environment. Furthermore, human behavior is governed not only by biological necessities but also by the desire for change. When surfeited with honey man begins to loathe the taste of sweetness, and this desire for change per se introduces an inescapable component of unpredictability in his life.

It is the awareness of these complexities which accounts for the clumsiness of the scientific language used in reporting biological events. The scientist emphasizes *ad nauseam* that what he states is valid only "under conditions of the experiment." As if apologetically, he is wont to qualify any assertion or general statement with the remark, "All other things being equal—which they never are . . ." Because things are never the same, almost everyone admits that prediction is always risky in political and social fields. But it is not so generally recognized that the same limitations apply to other areas usually regarded as falling within the realm of the so-called exact sciences, for instance, the epidemiology of disease.

Many examples have been quoted in earlier chapters to illustrate the unexpected and far-reaching effects that accidental circumstances have exerted in the past on the welfare of man. The introduction of inexpensive cotton undergarments easy to launder and of transparent glass that brought light into the most humble dwelling, contributed more to the control of infection than did all drugs and medical practices. On the other hand, a change in fur fashion brought about a few years later an outbreak of pneumonic plague in Manchuria; the use of soft coal in English grates caused chimney sweeps to develop cancer; Roentgen's discovery endangered the lives of scientists and physicians exposed to X rays in the course of their professional activities. Likewise oil and rubber may in the future come to be regarded as having been the indirect causes of disease and death. In addition to the human beings killed or maimed in automobile accidents, many are likely to suffer, directly or indirectly, from the air pollution brought about by the widespread use of oil and rubber. Furthermore, neuroses peculiar to our time may someday be traced to the speed and power that rubber and oil have made possible, as well as to the frustrations caused by crowded city streets and highways.

Human goals, which condition social changes, profoundly affect the physical and mental well-being of man. And, unfortunately, the most worth-while goals may have results as disastrous as those of the most despicable ambitions. Industrial imperialism was responsible for an enormous amount of misery among children during the early nineteenth century. But, as we have seen, the present philosophy to assure the survival of all children and to protect them from any traumatic experience also is likely to have unfortunate consequences by interfering with the normal play of adaptive processes.

Philosophical and social doctrines have been the most influential forces in changing the human ways of life during historical times. The high regard in which the human body was held by the Greco-Roman world certainly played a role in the development of hygiene and medicine during the classical times of Western civilization. In contrast, the emphasis on mystical values and on eternal life, the contempt for bodily functions, which characterized certain early phases of the Christian faith, probably led to the neglect of sanitary practices during medieval times—even though it did not necessarily decrease the enjoyment of sensual pleasures by normal men and women. Today, as in the past, the relation that man bears to his total environment is influenced by values of which he is not always aware. A civilization that devotes page after page of its popular magazines to portraying the rulers of the business world is bound to produce men very different from those taught to worship Confucian wisdom, Buddhic mysticism, or Blake's poems—even if that worship often does not go far beyond mere lip service. To feel at ease among the neon lights of Broadway demands a type of body and mind not conducive to happiness in the mists of a Taoist moonscape.

Technology is now displacing philosophical and religious values as the dominant force in shaping the world, and therefore in determining human fate. What man does today and will do tomorrow is determined to a large extent by the techniques that expert knowledge puts at his disposal, and his dreams for the future reflect the achievements and promises of the scientists. From them he has acquired the faith—or rather the illusion—that society can be planned in a manner that will assure plenty, health, and happiness for everyone and thus solve all the great problems of existence.

As modern technological innovations are the direct outcome of scientific research, scientists can no longer afford to stand aloof from social problems. Knowledge can grow without regard for ethical values, but the modern scientist cannot help becoming involved in ethics, since science can no longer be dissociated from the applications of science. In the past the social effects of science were slow in manifesting themselves. Today they are immediate and reach every aspect of the life of every man, for good and for evil.

The scientist has convinced society that his efforts deserve to be generously supported because he has become one of its most effective servants. As a penalty for his dependence on public support and for the influence that he has gained he cannot escape being made responsible for his activities, even if their results are different from what he had hoped. In the present decade he has to deal with the consequences of the release of man-made radiations. He may soon acquire the knowledge that will permit him to control the behavior of people and the genetic endowment of children to be born, a power frightening in its unpredictable potentialities for evil.

To discover, to describe, to classify, to invent, has been the traditional task of the scientist until this century, on the whole a pleasant occupation amounting to a sophisticated hobby. This happy phase of social irresponsibility is now over and the scientist will be called to account for the long-term consequences of his acts. His dilemma is and will remain that he cannot predict these consequences because they depend on many factors outside his knowledge or at least beyond his control—in particular on the exercise of free will by men. The scientist must therefore avoid pride of intellect and guard himself against any illusion or pretense as to the extent and depth of what he knows. He must also develop an alertness to the unexpected, an awareness of the fact that many surprising effects are likely to result from even trivial disturbances of ecological equilibria. Fortunately, the scientific method is well suited for the cultivation of this alertness to the advent of the unpredictable. The scientist cannot predict the remote consequences of his activities, but he can often provide techniques for recognizing them early. One of the few encouraging indications that science has come of age is the fact that extensive studies on the potential danger of radiations were initiated as soon as it became apparent that the forces unleashed by knowledge of the atom would find a place in the technology of war and peace.

To become worthy of his power the scientist will need to develop enough wisdom and humane understanding to recognize that the acquisition of knowledge is intricately interwoven with the pursuit of goals. It has often been pointed out that the nineteenth-century slogan, "Survival of the fittest," begged the question because it did not state what fitness was for. Likewise it is not possible to plan man's future without deciding beforehand what he should be fitted for, in other words, what human destiny ought to be —a decision loaded with ethical values. What is new is not necessarily good, and all changes, even those apparently the most desirable, are always fraught with unpredictable consequences. The scientist must beware of having to admit, like Captain Ahab in Melville's *Moby Dick*, "All my means are sane; my motives and objects mad."

Health, Happiness, and Human Values

It is often suggested that a moratorium on science would give mankind the opportunity to search its soul and discover a solution to the problems that threaten its very survival. Although no one is naive enough to hope that stopping the clock would bring about the solution of ancient human problems, many believe that a scientific status quo might prevent or retard the development of new threats. This static formula of survival is not new; indeed, it has been used with much biological success by social insects. Certain species of ants and termites had completed at least fifty million years ago the highly stratified and efficient type of colonial organization which they still exhibit. They have solved many of the problems which are the subject of endless discussions and conflicts in most human societies. Their queens, warriors, and workers all are produced as needed by genetic and physiological control, they have functions which are clearly defined and regulated in terms of the welfare of the colony as a whole. Even problems of eugenics have been solved in these insect societies by confining reproduction to a certain caste and promptly eliminating all abnormal and diseased individuals.

The very survival and wide distribution of highly organized insect societies which have not changed in fifty million years is evidence that living things can achieve a more or less stable equilibrium with their environment and that, beyond a certain degree of adaptation, change is no longer necessary for biological survival. It is conceivable, therefore, that human societies also could stop evolving and thus avoid the dangers inevitably associated with the adaptive problems bound to arise from any change. In fact, this has happened on several occasions in many parts of the world.

Before their contact with the white man the Eskimos, the Polynesian Islanders, and certain nomadic tribes had worked out stable societies with an acceptable degree of physical health and happiness. As pointed out by Arnold J. Toynbee, however, the human beings in all these societies were degraded by specialization and by limitation of their activities to a level far below that of the ideal all-round men evoked in Pericles' funeral speech. These "arrested" societies resembled in some respects the societies of bees and ants. Their stability may have resulted in the avoidance of many new adaptation problems but proved incompatible with the growth of their civilizations, indeed, with the very growth of man. It was the awareness of this limitation which had estranged D. H. Lawrence from the Polynesian Paradise:

> There they are, these South Sea Islanders, beautiful big men with their golden limbs and their laughing, graceful laziness. . . . They are like children, they are generous: but they are more than this. They are far off, and in their eyes is an early darkness of the soft, uncreate past. . . . There is his woman, with her knotted hair and her dark, inchoate, slightly sardonic eyes. . . . She has soft warm flesh, like warm mud. Nearer the reptile, the Saurian age. . . .
>
> Far be it from me to assume any "white" superiority. It seems to me, that in living so far, through all our bitter centuries of civilization, we have still been living onwards, forwards. . . . The past, the Golden Age of the past—what a nostalgia we all feel for it. Yet we don't want it when we get it. Try the South Seas.

The fact that, except for a few arrested societies, man has been living and struggling forward in a great life-development shows that utopias and all static formulas of society are out of tune with the human condition. It is the desire for change which has set man apart from the rest of the living world, by leading him to a life of adventure away from the environments to which he was biologically adapted, and it is this desire that will continue to generate the creative forces of his future. The Athenians symbolize for us the most brilliant achievement of mankind because, according to Thucydides, "They go on working away in hardship and danger all the days of their lives, seldom enjoying their possessions as they are always adding to them. They prefer hardship and activity to peace and quiet."

Once his essential biological needs are satisfied, man develops other urges which have little bearing on his survival as a species. When he no longer needs to struggle for his loaf of bread he is wont to crave an unessential savory, then to long for some artistic expression. When he has established all kinds of direct and indirect contacts with the surrounding world he begins to worry about the next television set and soon longs to explore the rest of the universe. Indeed, it is probably the most distinguishing aspect of human life that it converts essential biological urges and functions into activities which have lost their original significance and purpose. Eating habits are now determined by acquired tastes and by social conventions rather than by nutritional requirements. The acts of love are performed for pleasure rather than for reproduction. "If all our women were to become as beautiful as the Venus de' Medici," wrote Charles Darwin in Chapter XIX of *The Descent of Man and Selection in Relation to Sex*, "we should be for a time charmed; but we should soon wish for variety, and as soon as we had obtained variety, we should wish to see certain characters a little exaggerated." Thus, man desires change for change's sake, without regard to any biological need. This desire expresses itself in the most ordinary manifestations of life, like the choice of food, and in the most sophisticated occupations, like the various forms of art. It

affects the newest technological developments, like the
hoods of motorcars, as well as the most ancient occupations, like hunting. Now that high-power rifles are available, sportsmen are returning to the use of primitive weapons.
In 1957 forty thousand adults registered for the right to
hunt with bow and arrow in the state of Michigan alone.

It is important, indeed, that there be available opportuni-
ties for change, for when they are lacking man is apt to
satisfy his thirst for change by acts of violence or destruction. Dostoevsky's sniveling hero in *Letters from the Under-
world* could not find satisfaction in the order and comfort
of the "Crystal Palace" world in which he lived; he chose
an antisocial way of life because it was the one form of
freedom of action still available to him. "Well, gentlemen,
what about giving all this commonsense a mighty kick . . .
simply to send all these logarithms to the devil so that we
can again live according to our foolish will?" "Man only
exists for the purpose of proving to himself that he is a man
and not an organ-stop! He will prove it even if it means
physical suffering, even if it means turning his back on civilization." Many forms of delinquency among our overfed
teenagers probably come from their unspent creative
energy.

Mankind behaves like the restless, sleepless traveler who
turns in his berth to one side and then to the other, feeling
better while changing position even though he knows that
the change will not bring him lasting comfort. This restlessness is commonly identified with the concept of progress.
In reality, however, the only certain fact is that human history is increasingly governed by the search for variety, at
times for the sake of creation, more commonly just for
recreation, but in any case unrelated to the forces which
determine the evolution of biological traits. Progress means
only movement without implying any clear statement of direction. At most it can be said that, despite so many disheartening setbacks, the activities of man seem to have on
the whole a direction upward and forward which tends to
better his life physically, intellectually, and morally.

The desire for progress may be nothing more than man's

declaration of independence from the blind forces of nature. To paint the Last Supper, to write a poem, or to build an empire demands the expenditure of a form of energy and produces a type of result which does not have an obvious place in the natural order of things. In fact, as we have seen, certain of man's ideals and goals threaten to have consequences unfavorable for the human species. The cultivation of refined or esoteric tastes may interfere with the play of adaptive mechanisms and render man more vulnerable to some of his ancient plagues. The very mastery of nature may release dangers that cannot be controlled. Changes in the social order which increase the richness and variety of life can also, especially if too rapid, upset the ecological equilibria on which depends the continuation of the human species.

Awareness of dangers is not likely to deflect the course of mankind, for man does not live by bread alone. "All man wants," wrote Dostoevsky, "is an absolutely *free* choice, however dear that freedom may cost him and wherever it may lead him." True enough, most men run almost mechanically like clocks from their birth to their death, motivated only by their biological needs of the moment and by the desire to feel socially secure. But their very passivity makes them of little importance for social evolution. The aspect of human nature which is significant because unique is that certain men have goals which transcend biological purpose.

Among other living things, it is man's dignity to value certain ideals above comfort, and even above life. This human trait makes of medicine a philosophy that goes beyond exact medical sciences, because it must encompass not only man as a living machine but also the collective aspirations of mankind A perfect policy of public health could be conceived for colonies of social ants or bees whose habits have become stabilized by instincts. Likewise it would be possible to devise for a herd of cows an ideal system of husbandry with the proper combination of stables and pastures. But, unless men become robots, no formula can ever give them permanently the health and happiness symbolized by the contented cow, nor can their societies achieve a structure that will last for millennia. As long as mankind is made up of independent individuals with free will, there cannot be any social status quo. Men will develop new urges, and these will give rise to new problems, which will require ever new solutions. Human life implies adventure, and there is no adventure without struggles and dangers.

Envoi

Men naturally desire health and happiness. For some of them, however, perhaps for all, these words have implications that transcend ordinary biological concepts. The kind of health that men desire most is not necessarily a state in which they experience physical vigor and a sense of well-being, not even one giving them a long life. It is, instead, the condition best suited to reach goals that each individual formulates for himself. Usually these goals bear no relation to biological necessity; at times, indeed, they are antithetic to biological usefulness. More often than not the pursuit of health and happiness is guided by urges which are social rather than biological, urges which are so peculiar to men as to be meaningless for other living things because they are of no importance for the survival of the individual or of the species.

The satisfactions which men crave most, and the sufferings which scar their lives most deeply, have determinants which do not all reside in the flesh or in the reasonable faculties and are not completely accounted for by scientific laws.

"Reason," wrote Dostoevsky, "can only satisfy the reasoning ability of man, whereas volition is a manifestation of the whole of life. . . . Reason knows only what it has succeeded in getting to know . . . whereas human nature acts as a whole, with everything that is in it, consciously, and unconsciously, and though it may commit all sorts of absurdities, it persists." Exact sciences give correct answers to certain aspects of life problems, but very incomplete answers. It is important of course to count and measure what is countable and measurable, but the most precious values in human life are aspirations which laboratory experiments cannot yet reproduce. As Haeckel pointed out, Richtigkeit—correctness—is not sufficient to reach Wahrheit —the real truth.

Homo sapiens as a biological machine may not have changed much since Pleistocene times, but mankind has continued to evolve, developing a new kind of life almost transcendental to its earthly biological origin. It is a paradoxical attribute of many human beings that their behavior is often governed by criteria and desires that they value more than life itself. To comprehend the biology of mankind, the story of human evolution, it is helpful to remember Aristotle's saying· "The nature of man is not what he is born as, but what he is born for." Indeed, some men in all ages have been guided by the faith that "he who would save his life first must lose it." Alone among living things, men are willing to sacrifice the purely biological manifestation of their existence at the altar of a higher form of life —conceived in the soul rather than experienced in the flesh. Even the least religious of thinking men believes in the deep symbolism of what Paul wrote of human nature: "It is sown a natural body; it is raised a spiritual body. . . . The first man is of the earth, earthy: the second man is the Lord from heaven."

Because man is a spiritual body he is more concerned with a way of life than with his physical state. Balzac, on his deathbed, projected Herculean labors and pleaded with his physician to keep him alive six weeks longer in order that he might finish his work. "Six weeks with fever is an eternity. Hours are like days . . . and then the nights are not lost." Marcel Proust, also on the day before he died, wrote of those obligations of the artist which seem to be derived from some other world, "based on goodness, scrupulousness, sacrifice."

"Work is more important than life," Katherine Mansfield confided to the last pages of her *Journal*. Searching for a definition of health that would satisfy her body riddled with tuberculosis and also her tormented soul, she could only conclude, "By health, I mean the power to live a full, adult, living, breathing life in close contact with what I love—the earth and the wonders thereof—the sea—the sun. . . . *I want to be all that I am capable of becoming*, so that I may be . . . there's only one phrase that will do—*a child of the sun.*"

The sun is not merely a source of warmth, of light, of food, of power. It is also the symbol of human aspirations. Like Icarus, who, soaring upward to heaven, plummeted to the sea and died when his waxen wings were melted by the sun, man deliberately exposes himself to dangers and even to destruction whenever he tries to escape from his biological and earthly bondage. Wherever he goes, whatever he undertakes, he will encounter new challenges and new threats to his welfare. Attempts at adaptation will demand efforts, and these efforts will often result in failure, partial or total, temporary or permanent. Disease will remain an inescapable manifestation of his struggles.

While it may be comforting to imagine a life free of stresses and strains in a carefree world, this will remain an idle dream. Man cannot hope to find another Paradise on earth, because paradise is a static concept while human life is a dynamic process. Man could escape danger only by renouncing adventure, by abandoning that which has given to the human condition its unique character and genius among the rest of living things Since the days of the cave man, the earth has never been a Garden of Eden, but a Valley of Decision where resilience is essential to survival. The earth is not a resting place. Man has elected to fight, not necessarily for himself, but for a process of emotional, intellectual, and ethical growth that goes on forever. To grow in the midst of dangers is the fate of the human race, because it is the law of the spirit.

EPILOGUE

What "World Perspectives" Means

by Ruth Nanda Anshen

This is a reprint of Volume XXII of the WORLD PER-
SPECTIVES SERIES, which the present writer has
planned and edited in collaboration with a Board of Editors
consisting of NIELS BOHR, RICHARD COURANT, HU
SHIH, ERNEST JACKH, ROBERT M. MACIVER,
JACQUES MARITAIN, J. ROBERT OPPENHEIMER,
I. I RABI, SARVEPALLI RADHAKRISHNAN, ALEX-
ANDER SACHS.

This volume is part of a plan to present short books in
a variety of fields by the most responsible of contemporary
thinkers. The purpose is to reveal basic new trends in modern civilization, to interpret the creative forces at work in
the East as well as in the West, and to point to the new
consciousness which can contribute to the universe, the individual and society, and of the values shared by all people. *World Perspectives* represents the world community of
ideas in a universe of discourse, emphasizing the principle
of unity in mankind of permanence within change.

Recent developments in many fields of thought have
opened unsuspected prospects for a deeper understanding
of man's situation and for a proper appreciation of human
values and human aspirations. These prospects, though the
outcome of purely specialized studies in limited fields, require for their analysis and synthesis a new structure and
frame in which they can be explored, enriched, and ad-

vanced in all their aspects for the benefit of man and society. Such a structure and frame it is the endeavor of *World Perspectives* to define, leading hopefully to a doctrine of man.

A further purpose of this Series is to attempt to overcome a principal ailment of humanity, namely, the effects of the atomization of knowledge produced by the overwhelming accretion of facts which science has created: to clarify and synthesize ideas through the *depth* fertilization of minds; to show from diverse and important points of view the correlation of ideas, facts, and values which are in perpetual interplay; to demonstrate the character, kinship, logic, and operation of the entire organism of reality while showing the persistent interrelationship of the processes of the human mind and in the interstices of knowledge; to reveal the inner synthesis and organic unity of life itself.

It is the thesis of *World Perspectives* that in spite of the differences and diversity of the disciplines represented, there exists a strong common agreement among the authors concerning the overwhelming need for counterbalancing the multitude of compelling scientific activities and investigations of objective phenomena from physics to metaphysics, history and biology, and to relate these to meaningful experience. To provide this balance, it is necessary to stimulate an awareness of the basic fact that ultimately the individual human personality must tie all the loose ends together into an organic whole, must relate himself to himself, to mankind and society, while deepening and enhancing his communion with the universe. To anchor this spirit and to impress it on the intellectual and spiritual life of humanity, on thinkers and doers alike, is indeed an enormous challenge which cannot be left entirely either to natural science on the one hand nor to organized religion on the other. For we are confronted with the unbending necessity to discover a principle of differentiation yet relatedness lucid enough to justify and purify scientific, philosophic, and all other knowledge while accepting their mutual interdependence. This is the crisis in consciousness made articulate through the crisis in science. This is the new awakening.

World Perspectives is dedicated to the task of showing that basic theoretical knowledge is related to the dynamic content of the wholeness of life. It is dedicated to the new synthesis at once cognitive and intuitive. It is concerned with the unity and continuity of knowledge in relation to man's nature and his understanding, a task for the synthetic imagination and its unifying vistas. Man's situation is new and his response must be new. For the nature of man is knowable in many different ways and all of these paths of knowledge are interconnectable and some are interconnected, like a great network, a great network of people, between ideas, between systems of knowledge, a rationalized kind of structure which is human culture and human society. Knowledge, it is shown in these volumes, no longer consists in a manipulation of man and nature as opposite forces, nor in the reduction of data to statistical order, but is a means of liberating mankind from the destructive power of fear, pointing the way toward the goal of the rehabilitation of the human will and the rebirth of faith and confidence in the human person. The works published also endeavor to reveal that the cry for patterns, systems, and authorities is growing less insistent as the desire grows stronger in both East and West for the recovery of a dignity, integrity, and self-realization which are the inalienable rights of man who is not a mere *tabula rasa* on which anything may be arbitrarily imprinted by external circumstance but who possesses the unique potentiality of free creativity. Man is differentiated from other forms of life in that he may guide change by means of conscious purpose in the light of rational experience.

World Perspectives is planned to gain insight into the meaning of man who not only is determined by history but who also determines history. History is to be understood as concerned not only with the life of man on this planet but as including also such cosmic influences as inter-

penetrate our human world This generation is discovering
that history does not conform to the social optimism of
modern civilization and that the organization of human
communities and the establishment of freedom, justice, and
peace are not only intellectual achievements but spiritual
and moral achievements as well, demanding a cherishing
of the wholeness of human personality, the "unmediated
wholeness of feeling and thought," and constituting a
never-ending challenge to man, emerging from the abyss
of meaninglessness and suffering, to be renewed and re-
plenished in the totality of his life.

World Perspectives is committed to the recognition that
all great changes are preceded by a vigorous intellectual
re-evaluation and reorganization. Our authors are aware
that the sin of hybris may be avoided by showing that the
creative process itself is not a free activity if by free we
mean arbitrary or unrelated to cosmic law. For the creative
process in the human mind, the developmental process in
organic nature, and the basic laws of the inorganic realm
may be but varied expressions of a universal formative
process. Thus *World Perspectives* hopes to show that although the present apocalyptic period is one of exceptional
tensions, there is also an exceptional movement at work toward a compensating unity which cannot obliterate the
ultimate moral power pervading the universe, that very
power on which all human effort must at last depend. In
this way, we may come to understand that there exists an
independence of spiritual and mental growth which though
conditioned by circumstances is never determined by circumstances. In this way the great plethora of human
knowledge may be correlated with an insight into the nature of human nature by being attuned to the wide and
deep range of human thought and human experience. For
what is lacking is not the knowledge of the structure of the
universe but a consciousness of the qualitative uniqueness
of human life.

And, finally, it is the thesis of this Series that man is in
the process of developing a new awareness which, in spite

of his apparent spiritual and moral captivity, can eventually lift the human race above and beyond the fear, ignorance, brutality, and isolation which beset it today. It is to this nascent consciousness, to this concept of man born out of a fresh vision of reality, that *World Perspectives* is dedicated.

ANCHOR BOOKS

Lightning Source UK Ltd.
Milton Keynes UK
UKHW010726140223
416966UK00007B/128

9 781258 815714